The Energy Wise
Workplace

The Energy Wise Workplace

Practical and Cost-Effective Ideas for a Sustainable and Green Workplace

Jeff Dondero

ROWMAN & LITTLEFIELD
Lanham • Boulder • New York • London

Published by Rowman & Littlefield
A wholly owned subsidiary of The Rowman & Littlefield Publishing Group, Inc.
4501 Forbes Boulevard, Suite 200, Lanham, Maryland 20706
www.rowman.com

Unit A, Whitacre Mews, 26-34 Stannary Street, London SE11 4AB

British Library Cataloguing in Publication Information Available

Library of Congress Cataloging-in-Publication Data

Names: Dondero, Jeff, 1947– author.
Title: The energy wise workplace : practical and cost-effective ideas for a sustainable and green workplace / Jeff Dondero.
Description: Lanham : Rowman & Littlefield, [2017] | Includes bibliographical references and index.
Identifiers: LCCN 2016054646 (print) | LCCN 2017013292 (ebook) | ISBN 9781442279506 (ebook) | ISBN 9781442279490 (cloth)
Subjects: LCSH: Management—Environmental aspects. | Offices—Energy conservation. | Office buildings—Energy conservation.
Classification: LCC HD30.255 (ebook) LCC HD30.255.D66 2017 (print) | DDC 759/26—dc23
LC record available at https://lccn.loc.gov/2016054646

♾™ The paper used in this publication meets the minimum requirements of American National Standard for Information Sciences—Permanence of Paper for Printed Library Materials, ANSI/NISO Z39.48-1992.

Printed in the United States of America

Contents

Acknowledgments and Disclaimer

Thanks as always to Alicia, from whence all good things come. Kudos to my editor, Kathryn Knigge, who wades through copy both thick and thin with tenacity, care, and dedication; to my steadfast friend and colleague, Patrick Totty; and to Rowman & Littlefield for publishing *The Energy Wise Workplace*.

A remark generally attributed to Mark Twain applies here: "There are lies, damn lies, and statistics." Most of the stats here have been gathered by local, state, and federal government agencies, private parties, and other sources. Consequently, I do not claim that all of the information presented herein represents accurate and true statements, percentages, and facts, and I do not warrant or make any representations as to the content, accuracy, or completeness of the information, text, graphics, charts, Web links, Web sites, and other items contained in their media presentations.

Writing and aggregating information for this kind of book has its inherent dilemmas and predicaments. When presented with a question, people find different ways to get answers and draw conclusions. Consequently answers can vary, sometimes quite a bit. For example, prices for fuel and/or power vary not only between different utility companies, but also from county to county, state to state, and can fluctuate due to availability and demand.

Although most of the facts presented herein are defensible, they are used as literary and entertainment devices to give the reader a general and generic perspective on information and suggestions on how to save money and conserve valuable resources as well as the physical and psychological welfare of employees.

In an effort to communicate more easily and effectively, the author has sometimes taken averages, mean numbers, or common sense, and has modified statements to reflect more than one set of facts or opinions. Also, the results of certain criteria can be altered by many factors, including the size of a company, the number of employees, the nature of a business, and the various local and federal rules and regulations. In any case, the information presented herein is for entertainment as well as educational purposes and should be used in that spirit.

The lawyers put it another way: Neither the author, the publisher, nor any of their employees makes any warranty, or guarantee, express or implied, or assumes any legal liability or responsibility for the accuracy, completeness, or usefulness of any information, percentage, apparatus, product, device, or process disclosed, or represents that its use would not infringe privately owned rights. Reference herein to any specific commercial copyright, product, process, or service by trade name, trademark, manufacturer, or otherwise, does not necessarily constitute or imply its endorsement or recommendation. The views and opinions of the author expressed herein do not necessarily state or reflect those of Rowman & Littlefield Publishing or its agents. This book is for entertainment and educational purposes only.

How Green Is Your Workplace?

Aids for the Office

Much of the energy that is consumed to run a business comes from a nozzle, a faucet, a socket, or a switch. That energy runs machines, but the energy at the heart of a successful company comes from its employees. Both are necessary for sustainability. Simply defined, a sustainable system is one which operates in a way that it does not use up resources more quickly than they can be naturally replenished. You'll be reading a lot about not only how to conserve power and materials, but how to maintain the sustainability of one of your most important sources of energy—your people.

WHO MAKES THE DECISION TO
SAVE ENERGY AT YOUR COMPANY?

If energy efficiency is not an accepted philosophy of your business, the effect is a lot of wasted time, talent, and calories. It's very interesting to watch how the decision to save energy is made by whom and where. Getting someone to dive into that process and make decisions and face the challenges of implementation are two key factors as to why 30 percent of the energy used in businesses is typically wasted.

In many businesses the decision to save energy is considered the duty of the maintenance department, a business manager, or perhaps a sustainability specialist. The executive team is often more concerned with improving the bottom line through increased production or sales. Suggestions and feedback from the entire crew are important, especially when workers act as a team and really care about energy management and conservation. Unless it is an organization-wide philosophy that is fostered, apathy and energy waste are often the outcome.

Typically, things like lights and equipment get left on; heating, ventilation, and cooling (HVAC) systems become outdated; machine maintenance is not scheduled; and employees are not properly motivated. However, when saving energy is a company-wide value and management appropriately motivates and rewards the staff, every person who shuts off an unused light, lowers the thermostat, or reduces or recycles waste is making the decision to be more energy-efficient, which will show up on the bottom line.

GREEN-CHARGED WORKERS IN DEMAND

The demand for workers that mix green-thinking with core business skills is part of the new holistic corporate philosophy. Heavy hitters like Coca-Cola, Del Monte, and Toyota are looking for people skilled in the growing number of graduate business programs such as carbon accounting, sustainability related job skills, corporate social responsibility, and lean manufacturing techniques to reduce waste and environmental impact, while looking at ways to improve the triple bottom line: people, planet, and profits.

In order to jumpstart long overdue growth in clean power and policies, some of the world's largest companies, from Google to Staples, Microsoft to Walmart are making direct investments in renewable energy generation. The trend of large global companies supporting on-site installations of significant energy-generating capacity is one that is accelerating. Commercial buildings account for 19 percent of the energy consumed in the United States. The Department of Energy (DOE) expects that by 2030 office energy demand will grow by 25 percent. As a result of rising utility bills, direct investment in clean power promises financial returns that are comparable to—if not higher than—other successful investment options and is starting to look more and more like a smart bottom line decision.

WORKERS ARE TRENDING GREEN

Three-quarters of US workers think it's important that their employer take action to protect the environment. And one in four employees would likely look for a new job if they discovered their employer had a bad record on environmental issues, while more than 20 percent might accept a 5 percent pay cut to work for a company that takes strong environmental protection action.

Many of us spend almost as many waking hours at work as we do at home. In fact, the people who put in the largest number of working hours—contrary to bigoted belief—are Mexicans. The United States ranks around sixteenth in hours worked, according to *Forbes* magazine, and seventeenth in average income.

Unfortunately, we rank low in terms of employee and environmental stewardship, mainly because of our failure to keep up with most industrialized nations in caring for employees and their families, and because we sometimes injudiciously and voraciously prospect for and exploit natural resources such as land, water, air, and fossil fuels.

BUILDINGS ARE TRENDING GREEN TOO

There are more than 30 million workers employed in US office buildings working in more than 81 billion square feet of commercial floor space in 750,000 office and industrial buildings. Laid out on a single level, these buildings would cover Rhode Island two-and-a-half times.

Some of the energy-saving and conservation ideas that we practice at home aren't possible at the office, others are the province of building owners or superintendents. Employees don't have to strap on a tool belt or arm themselves with caulk guns to track down drafts and worn weather-stripping, but this book will look into many things each person can do to contribute to better conservation with all types of energy, including their own.

There are many suggestions that can help conserve resources, aid the environment, save your company money, engage your energy more efficiently, and help in reducing a large carbon footprint on the Earth.

According to the World Business Council for Sustainable Development (WBCSD), "Energy efficiency is fast becoming one of the defining issues of our time." The WBCSD invested four years and $15 million in what's being called the most rigorous study on energy efficiency in buildings ever conducted. According to the report, almost half the energy used in buildings is wasted, and energy use in buildings can be cut by 60 percent by 2050, but planning and action is necessary now.

The US DOE makes practical suggestions about building energy performance through the development and promotion of efficient, affordable, and high-impact technologies, devices, systems, and practices. The long-term goal of the Building Technologies Office, a DOE network of research and industry partners developing innovative, cost-effective, energy-saving solutions for homes and buildings, is to reduce energy use by 50 percent.

A snap term, "smart buildings," includes buildings that achieve significant energy and resource savings by taking advantage of automated, integrated technology for energy and management systems, and use materials that are healthy, harmonious, environmentally safe, reliable, and sustainable, featuring people-friendly systems that create a more productive atmosphere.

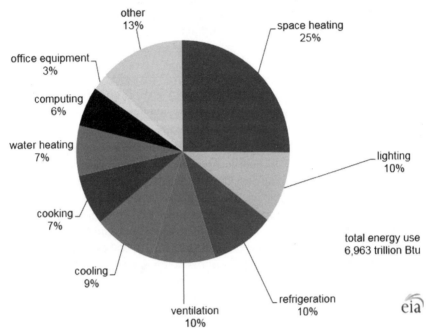

Figure 1.1.   Office Energy Use. Source: US Energy Information Administration (EIA).

HOW MUCH ENERGY AT WHAT COST?

By 2025, buildings are predicted to be the largest consumers of energy, more than transportation and the industrial sector combined. In California alone, businesses collectively spend more than $15 billion a year on heating, cooling, lighting, and other office energy uses. Office buildings in the United States consume an average of $1.34 per square foot of electricity and $0.18 on natural gas.

Electricity consumption in buildings doubled between 1989 and 2005. If this growth rate is sustained, electricity demand in buildings may increase another 50 to 100 percent by 2030, as new electronic gadgets and devices increase in our workplace.

ARE WE EFFICIENT?

On average, EnergyStar buildings (government-backed rating program that helps people and organizations save money and reduce greenhouse gas emissions) use 35 percent less energy than average buildings and cost $0.54 less per square foot to operate when compared to an average efficient building, and $2.10 less per square foot to operate than inefficient buildings.

Renovations and retrofits that replace older systems with more ecologically sound and cost-effective technology in existing buildings can reduce businesses' energy costs by as much as 30 percent. Also, a pleasant work environment may actually help attract the best and the brightest worker since up to 90 percent of an office worker's time is spent indoors. Studies have shown that "smarter" buildings have higher occupancy rates, higher productivity, and are more economical and efficient than their competitors by about 10 percent.

US office buildings have failed to keep pace with the energy conservation revolution in automation, according to IBM. Nationwide, only 33 percent of office workers rated their office buildings "somewhat high," "very high," or "extremely high" in terms of environmental responsibility. Seventy percent of those surveyed said that they make an effort to conserve resources such as water or electricity as part of their regular routine at work. Seventy-five percent say they would be more likely to conserve resources at work if they were rewarded for the effort, 65 percent say they would participate in the redesign of the workspace in their office buildings to make them more environmentally responsible, and 14 percent report that their office buildings make use of solar energy or another renewable energy source.

Although urinals account for up to 20 percent of a commercial building's water use, only 31 percent state their office buildings have low-flow toilets. A single waterless urinal can save almost 40,000 gallons per year.

Heating and cooling is the single most-expensive cost in a building. A simple change in roof color (white) can often reduce cooling demand by 15 to 20 percent.

EQUIPMENT AND ENERGY

Participants in the EPA Green Lights program (an innovative program that encourages US corporations to install energy-efficient lighting technologies) are using lighting options to cut their power bills in half and earn an average 58 percent return on their investment (http://www.dazor.com/green-lights.html).

The EnergyStar Challenge is to improve and set goals for energy use by committing to a 10 percent reduction of energy use in office buildings. The

Carbon Trust estimates that up to 15 percent of the energy consumed by today's offices can be attributed to idle office equipment.

THE "OFF" BUTTON IS A SIMPLE SOLUTION

Computers in the office burn $1 billion worth of standby electricity (when they're not being used) in the United States annually. Significant energy savings can be found in most office equipment, which is inactive more often than it is active. According to the DOE, annual energy use by personal computers is expected to grow 3 percent per year, and energy use among other types of office equipment is expected to grow more than 5 percent. If every US computer and monitor were shut off every night, eight large power stations could be shut down and carbon dioxide emissions would decrease by 7 million tons annually. Turn-off rates are highest, 40 percent, among multi- or integrated-computer systems, copiers, and scanners. The lowest turn-off rates, 20 percent, are among laser printers, LCD monitors, and multi-function devices (MFDs). Overall, 30 percent of energy consumed in buildings is used inefficiently or unnecessarily. Twenty billion dollars would be saved if the energy efficiency of commercial and industrial buildings improved by 10 percent.

In a small EnergyStar office, equipment can save more than 3,500 kilowatt-hours of electricity per year or about $265 dollars. The energy savings from 10 million office workers turning off unneeded lights for 30 minutes a day is enough to illuminate 1.3 million sq. ft. (four million square meters) of office space. According to a field survey by Lawrence Laboratories, the internal power management (PM) of specific devices is most successful among monitors and laser printers and least successful among desktop computers, inkjet printers, copiers, and fax machines.

Overall, the EPA estimates that the EnergyStar Office Equipment program will save 21 billion kilowatt-hours and 2.3 billion pounds of carbon dioxide a year—the equivalent of taking 807,000 cars off the road.

POLLUTION AND WASTE

In large urban areas like New York City approximately 80 percent of the carbon footprint comes from buildings. The EPA says that energy use in commercial buildings accounts for more than 12 percent of US greenhouse gas emissions at a cost of over $100 billion per year. Over the next twenty-five years, greenhouse gas emissions from buildings are projected to grow faster than those in any other sector, with emissions from commercial buildings leading the way—a growth of 2 percent a year through 2030. Buildings also are responsible for creating massive amounts of heat and can raise the temperature of a city by 3 percent (see Heat Island Effect).

Get off the mailing lists. Almost 50 percent of all catalogs are never opened, yet nearly 62 million trees are destroyed and 28 billion gallons of water are used to produce catalogs and brochures every year. Be attentive to water use. Up to 50 percent of water that flows into buildings today is wasted.

SAVE ENERGY SAVE $$

Rocky Mountain Institute, an organization dedicated to the field of sustainability, claims that in industrial settings there are abundant opportunities to save 70 to 90 percent of the energy costs for lighting, fans, and pump systems, 50 percent for electric motors, 60 percent in areas such as heating, cooling, office equipment, and appliances. In general, up to 75 percent of the electricity used in US businesses today could be saved with efficiency measures that cost less than the electricity itself.

According to the EPA's EnergyStar Challenge, improving the energy efficiency of US commercial and industrial buildings by just 10 percent would reduce greenhouse gases equal to the emissions from about 30 million vehicles. Los Angeles emerged as the clear winner in an IBM smarter buildings study, which surveyed 6,486 office workers in sixteen US cities on issues ranging from office building automation and security to elevator reliability and conservation issues. Early studies reveal that EnergyStar-labeled buildings on average consistently use 35 percent less energy than their peers and emit 35 percent less carbon dioxide.

Old is not aged. In 1999, the EPA awarded the EnergyStar to a seventeen-year-old, 74,000-square-foot rehabilitated municipal office building in San Diego, California, and to a church in Needham, Massachusetts, built in 1836. The oldest building awarded an EnergyStar is Cambridge Savings Bank, in Cambridge, Massachusetts, that was constructed in 1820, the same year in which Maine became the twenty-third state in the Union.

SIMPLE SUGGESTIONS

To begin with, introduce energy-efficiency programs like energy management system (EMS) and switch power management (PM) into temperature setback mode during unoccupied hours.

Join the EPA program called Green Lights where participants realize average rates of return on their initial investment of 30 percent or more. They reduce their lighting electricity bill by more than half while maintaining and often improving lighting quality.

Use everyday common sense like taking the stairs. It's good for your heart and saves electricity. Watch overtime—energy costs may go up in the evening and it costs a lot for only one or two people to work late in an office building. Carefully select location, traffic flow, and use of office equipment. Save energy by setting hibernation of computers after more than fifteen minutes of inactivity.

Use available natural light from skylights and windows and use creative office design, such as color and style, for aesthetics and reflective light. Reduce business travel by increasing phone, video, and Internet conferencing and training capabilities. Install low-flow toilets, urinals, faucets, and showerheads.

Regular examination of your building's maintenance systems can save from 6 to 20 percent in utility costs. Energy efficiency can be used as a low-risk, high-return investment: for every $1 invested in energy efficiency, asset value increases by an estimated $3. A 30 percent reduction in energy consumption can lower operating costs by $25,000 per year for every 50,000 square feet of office space. Save more than 30 percent of office equipment energy by using devices with the EnergyStar label and by using power strips that have a turn-off switch.

Estimate your savings by using an online energy savings calculator like the one at https://www.sba.gov/content/energy-saving-calculators-energy-star. And an overall energy audit would be a good investment for most businesses.

SUPPORT COMPANY AND EMPLOYEE INVOLVEMENT

Educate and encourage employees to be energy-conscious and to offer their ideas about how energy can be saved. Employee buy-in and involvement can make or break your company's efforts to conserve energy.

Designate a consultant or someone to be responsible for and to promote good energy practices for the organization and/or facility. This individual should work with management to facilitate energy-saving ideas and strategies, optimizing energy use and minimizing operation and overhead costs. Initiate prizes and praise for those offering energy-saving ideas.

Offer assistance for those taking courses in energy and resource efficiency. Make knowledgeable people available to speak at the office, attend, participate in, or host a booth at green events and conferences, or offer a "sustainability officer" to visit schools or local meetings as a guest speaker. Post job openings to Web sites with green job boards such as those hosted by JustMeans, treehugger, GreenBiz, Net Impact, SustainableBusiness, grist, and Monster. When offering internships, ask candidates about their concerns and what is important to them.

Encourage employees to use social and professional networking sites to connect with green recruiters. Socially conscious employees will be attracted to companies that support their desire to make positive contributions. Match employee contributions to social and environmental causes and offer time off for employees who wish to speak on green issues.

ENERGY SAVING RESOURCES AND AIDS

- National Grid offers technical assistance and financial incentives to help purchase and install energy-efficient equipment for your facility.
- Find energy-saving product recommendations and spreadsheets to estimate your savings.
- The Building Owners and Managers Association (BOMA) Energy Efficiency Program (BEEP) offers online training on low-cost adjustments for energy savings, and estimates of financial returns.
- The Web sites dealing with government benefits, grants, and financial aid may also be of help. Contact the federal Office of Energy Efficiency and Renewable Energy (Department of Energy–DOE), 528 Senate Hart Office Building, Washington, DC 20510 (202-224-4744 begin_of_the_ skype_highlighting).
- Join the Advanced RTU (Rooftop Unit) or ARC Campaign to realize 20 to 50 percent energy savings.
- Check out the Lighting Energy Efficiency in Parking (LEEP) Campaign to lower your exterior lighting costs.
- Join the Wireless Meter Challenge, a cost-effective, wireless metering system capable of electrical energy measurement at various locations in a building and wireless communication to a remote data collection point within the building complex.
- The DOE's Building Technologies Office (BTO) works to identify and develop strategies and technologies to dramatically reduce commercial building energy consumption and is targeting a 20 percent energy use reduction in commercial buildings by 2020. To reach these goals, BTO engages building owners, builders, engineers, architects, contractors, manufacturers, and others to implement real-world energy-saving opportunities.

- DOE funded the Business Case for Energy Efficient Building Retrofit and Renovation Report, which provides insights on improving the energy consumption of existing buildings and developing the business case to seize these opportunities.
- Commercial Building Partnerships involve companies working with the DOE on specific retrofit or new construction projects in order to achieve whole-building energy savings.
- In the Assessment of the Technical Potential for Achieving Net Zero–Energy Buildings in the Commercial Sector report, engineers evaluated opportunities to significantly reduce energy consumption in construction of commercial buildings.
- The DOE's Better Buildings Alliance is an informal association of commercial building owners and operators working to develop resources and best practices to reduce energy consumption and carbon emissions in their industry.
- Check out your local utility companies for advice, energy-efficiency loans, retrocommissioning, and customized incentives.
- Contact the US Green Building Council, developers of Leadership in Energy and Environmental Design (LEED) for all manners of tips for conservation of resources and building upgrading and sustainability. LEED-certified buildings are resource-efficient, use less water and energy, and reduce greenhouse gas emissions.

2

The Office Power Struggle

The Energy Needs of Digital Devices

When people look back on the early days in a newspaper office compared to now, it's like being catapulted from the Stone Age into the Space Age. It seemed almost overnight that we went from manual typewriters and typesetters, Xacto razors, single-page copiers, waxers, blueline formatting paper, and "positive cameras" for graphic reproduction to computer terminals, pagination software, multi-use copier-scanners, and many types of software. While there's always some nostalgia for a simpler past, the sometimes aggravating automated age has given us tools that allow us to accomplish more than twice the work of one person which, for the most part, is a good thing.

These days it's becoming essential to watch the outlets, the power switches, and the various office appliances that eat up power like a pack of energy wolves. In a one-person office, there may be about a half-dozen multiple-use machines that are like little piggies fed by two or three power strips.

Back in the day, the power bill was less of a concern since much of the office equipment was manually operated. Climate change was noted by looking out the window and checking the weather, the talk about an excess of CO_2 in the atmosphere was primarily about smog, people washed their cars in the driveway and left the hose on, and there were only a few electrical appliances in the office.

Perhaps people were naïve or maybe they just weren't paying attention, but now our ignorance and excess are biting us in some soft places, like our wallets and environment.

You may have less control over power and resources at work than you do at home, but there is still plenty each person can do to mitigate the waste of resources, lower the company's utility bills (the boss will be grateful), and give a leg up to the planet—both locally and globally.

FACTS AND FIGURES

Always buy EnergyStar-qualified products for your small business. The EnergyStar mark indicates the most efficient computers, office equipment, refrigerators, televisions, windows, thermostats, ceiling fans, and other appliances. EnergyStar is a government-backed voluntary program that helps businesses and individuals save money and protect our climate; it has prevented more than 150 million metric tons of pollution from being put into the atmosphere in the last decade.

Monitoring your energy use and paying attention to the on-and-off switch can save you upwards of 6 percent of the utility bill. EnergyStar-labeled office equipment reduces energy use by power management (PM) that is automated to turn off or enter into a low power mode after fifteen to thirty minutes of inactivity. According to EnergyStar, a 10 percent decrease in energy use can lead to almost a 2 percent increase in net operating income. For a 200,000-square-foot office building that pays $2 per square foot in energy costs, a 10 percent reduction in energy consumption can translate into an additional $40,000 in net operating income.

Several surveys found that 90 percent of companies studied had energy management as a goal and more than two-thirds identified reducing energy costs as their primary motive. Yet, the surveys also found that very few companies have made significant energy efficiency improvements. Recent Energy Action (an energy software tool by Schneider Electric, specialists in energy management) assessments of ninety commercial buildings have identified cost recovery opportunities ranging from 20 to 40 percent.

Because of their large numbers, computers and monitors are the largest energy consumers of all office equipment. According to a report by Forrester Research, there are more than one billion PCs in use worldwide. And while it took it took twenty-seven years to reach the one billion mark, it will take only seven more to grow to two billion. A United Nations study found that the manufacturing of a computer and its screen takes at least 530 pounds of fossil fuels, 48 pounds of chemicals, and 1.5 tons of water. Most of the energy used during the life cycle of a computer with a 17-inch monitor is manufacturing the product, not computing.

Data centers (huge banks of servers and the air conditioning they need to keep them cool) can account for as much as 40 percent of an office's electricity bill. Switch off copiers and printers at the end of the working day and save 15 pounds of carbon dioxide emissions per week for a large copier, 5 pounds a week for a large printer.

PRIORITIZING ENERGY-EFFICIENCY PROJECTS

Common questions include where to start. Should one system be replaced, or would it be better to replace one piece of equipment at a time, or should a comprehensive upgrade of the entire facility be tackled? The answers will vary depending on each individual business situation. The age of the current equipment and facility systems, the type of business, local utility rates, hours of operation, and access to capital are all key factors as to what level of upgrade makes business sense.

If cash flow is an issue, you may want to wait until a piece of equipment or system fails or is a certain number of years old before replacing it with an energy-efficient model. However, if you are building a new facility or doing a major remodel, it makes more sense to incorporate energy-efficient upgrades into your design due to the lower incremental costs of "doing it right the first time." For an existing facility, it may come down to what is financially feasible for your business at a particular time.

OFFICE EQUIPMENT

Whether it's a car or a computer, people have passionate love-hate relationships with machines. On top of the list are those machines that help us run our offices, and on which we spend a lot of time and resources. Office equipment and other miscellaneous devices account for more than 20 percent of electricity consumption in large office buildings (counting some plug-in appliances like desk lamps and fans, etc.).

According to the US DOE, office equipment is the fastest-growing source of electricity consumption in businesses, including computers, monitors, printers, fax machines, and copiers. And energy use by office equipment is expected to grow by as much as 500 percent in the next decade (see figure 2.1).

Follow these tips to ensure your computers and related equipment are not using unnecessary power by turning off all equipment when not in use, especially overnight, which can cut annual energy costs by as much as $200 per computer, per year. Share printers by linking one or more to several computer stations. Consider laptops as they use about one-quarter the electricity of most desktop models.

EnergyStar-labeled monitors use up to 90 percent less energy than older models and those without power management features. Use smart power strips with built-in occupancy sensors that shut off plugged-in devices when the machines aren't being used after a short period of time, thereby ensuring that standby power use is stopped or reduced when power strips are turned off.

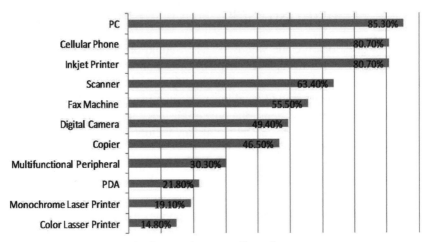

Figure 2.1.   Energy Use of Office Devices. By Jeff Dondero.

Use power management settings to modify energy settings within the Control Panel on Windows computers and in the System Preferences menu on Apple computers. This use of energy-efficient technologies can translate into reduced energy costs for computers in new buildings by 50 percent or more.

Overall, the EPA estimates that the EnergyStar Office Equipment program will save 3,500 kilowatt-hours of electricity per year and $265 for a medium-sized office, and 21 billion kilowatt-hours and 2.3 billion pounds of carbon dioxide a year—the equivalent of taking 807,000 cars off the road.

Smart meters track energy use and time-based rates to help avoid blackouts, aid in energy use, curb greenhouse gas emissions, and mitigate the need to build new power plants. Collect your utility bills. Separate electricity and fuel bills. Target the largest energy consumer or the largest bill for energy conservation measures.

If your budget allows, check out newer computer models. Flat screen monitors are becoming the most energy-efficient part of the system. Although they do cost more, they will save you money (about one-third less energy consumed) and office space.

Before putting in a purchase order request for the newest, fanciest model on the market, take a close look at exactly what programs and operating capacities employees need, and then try to upgrade and improve (instead of replace) wherever possible. Even small changes, like more memory and a new monitor, can make a big difference to the bottom line and can minimize trash build-up and cash outflow.

Clearly label any switches or equipment that must be left on, and then turn everything else off. Equipment only used occasionally should be unplugged

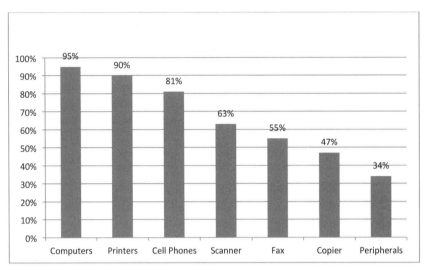

Figure 2.2. Daily Use of Office Machines. By Jeff Dondero.

altogether—items like DVD players and televisions, chargers, AC adapters, and other plug-ins that are rarely used.

Upgrade to an EnergyStar multi-function device (MFD) and toss out the separate fax machine, scanner, printer, and copier to reduce carbon dioxide emissions by more than 20 pounds per week and save office space. Multi-function devices also warm up quickly and use next to no standby power.

A single monitor draws about 100 watts per day; if left on overnight and on weekends it could cost more than $30 annually. Configure your monitor and computer to go to sleep (blank screen) after twenty minutes of inactivity. Use black and white monitors when possible as color monitors use twice as much energy.

Computers in sleep mode cut energy use down to 10 percent, which can save $25 to $75 per computer annually. If possible use laptops with 3.3-volt components (processor, memory, and LCD). These systems use 40 to 50 percent less energy than 5.0-volt systems, and are generally equipped with a lighter battery.

Other big energy users and greenhouse gas emitters in offices are the lights. Switching them off when they're not needed will save big time.

Use half the power for LCD monitors by setting them to run on low brightness (20 watts compared with 34 watts on full brightness). Flat-panel liquid crystal display (LCD) monitors use considerably less electricity than comparably-sized CRT models. There is a potential to save up to $178 a year by completely turning off each computer and monitor.

If you have to buy a CRT (cathode ray tube) monitor, buy the smallest monitor that will meet your needs. For example, a 17-inch monitor consumes 35 percent more electricity than a 14-inch monitor. Turning off one monitor could save approximately $90 on the annual energy bill. Turning off computers, monitors, printers, copiers, and lights during nonbusiness hours can save up to $178 a year. Add that figure up for multiple workstations in an office building, and wasted energy translates into a significant wad of wasted money.

Purchase the proper sized copier for your business needs. EnergyStar copiers are equipped with low-power mode features after fifteen minutes of inactivity and switch to an off mode of 5 to 20 watts after two hours of inactivity. This ensures that the equipment is off in the evening and on weekends, reducing annual copier electricity costs by up to 60 percent.

There is a 30 to 60 percent potential for savings in energy costs if you purchase new printers and copiers with fast warm-up times. According to EnergyStar, you should set your printer's low-power mode depending on the print speeds of your inkjet or laser printer. Economy mode uses up to 50 percent less toner and prints twice as many pages as higher quality settings. Avoid printing in color if possible. Use the duplex feature of your printer; it uses up less energy than single-sided copying.

Inkjet printers use 4 percent less and dot-matrix printers use 15 percent less energy than laser printers. An average device consumes around 20 percent of its full printing power when in standby. In contrast, sleep mode consumes only around 0.5 percent of full printing power.

When you must print a document, view your document first using your application's print preview feature prior to printing to avoid mistakes. Practice on-screen reading and editing habits. Add a line at the end of your employee email signatures that encourages people to consider the environment before printing any e-mail. Recycle your ink cartridges. Many office supply companies offer credit for recycling cartridges.

SERVERS

Servers chew up a lot of electricity, and they run all the time. Set the monitor to shut down when not needed, and set up power management features in the server's software. Turn off monitors when not in use.

When possible, consolidate multiple servers to a single, higher-performance device and utilize virtual machine technology to reduce the number of physical devices necessary. Think about utilizing server technology in data centers built to manage heating and cooling in a cost-effective manner.

COLLATERALS

Even though accessory devices might be a second thought in an office, they have big energy appetites. Use "smart" power or surge strips with switches for added convenience and power strips with built-in occupancy sensors that shut off plugged-in devices such as printers and monitors when machines are not in use.

Fax machines need to be available twenty-four hours a day but can be powered down to 15 to 45 watts or less when not in use, which can save over 50 percent on annual faxing energy costs.

Plasma TVs vary hugely in the amount of power they use—from 150 to 300 watts (producing 0.3 kg CO_2 per hour) for a 42-inch screen. Check manufacturer specifications or the specification plate on the back of the TV. Some modern plasma and LCD TVs actually sense the brightness of a room and will save energy by dimming the screen.

If you have a dishwasher in the office kitchen—an efficient one generates around 3 pounds of CO_2 per wash— switch off the electric hot water system and save 4 pounds of carbon dioxide per day from the hot water tank's heat loss. A modern dishwasher in an economy cycle (good enough for most office plates and cups) uses about half as much energy as a full cycle. Unless you're doing a hand-wash with light rinsing, you're better off using a dishwasher. And you'll never have to read one of those "Your mother doesn't work here/ clean your dishes" notices again. A bar fridge in the meeting room generates up to 1,320 pounds of CO_2 each year, depending on its efficiency. A refrigerated drink vending machine generates around 3 tons of greenhouse gas each year and about half of this is due to the display lights in it.

Microwaves with display clocks use more energy to power the clock than they do to heat up food or drink, so switch it off when not in use.

Look for coffeemakers, hotplates, and other appliances that can be switched off, after all leftover coffee doesn't need to be heated for hours.

Data projectors can use up to 300 watts of power (producing about 0.3 kg CO_2 per hour). Dimming or switching off lights when you're using a data projector can save as much energy as the projector uses.

FURNISHINGS

Buying furniture for the office should take time and energy. The choices you make will have a big impact on office environment, employee comfort, and overall productivity. Office workers who have choice and control over where they work are more engaged than those who don't. While open offices promote collaboration, there are times when employees need a more private space to focus on the task at hand or to make a quick phone call. Consider which different spaces your employees will need and benefit from when designing your workplace.

It's bottom-line tempting to purchase lower-priced furniture that may have the potential to cost you extra in terms of repairs and replacements. Office furniture is an investment, so think long-term. Buy furniture that can be reconfigured and moved easily when your staff and layout require changes.

Don't forget comfort, after all people will be using the furniture constantly and if they are not comfortable, they will get agitated which will undoubtedly impact productivity, performance, and workplace morale. Ergonomic chairs and desks (intended to provide optimum comfort and to avoid stress or injury) are critical as they make working easier and healthier, as supported by numerous studies. Ergonomics considerations like contoured seats, lumbar backrest supports, adjustable seats, and armrests are important.

Don't forget to account for the type of material used in your furniture, and select something that doesn't shows stains and is easy to clean. Choose materials and pieces that have consistency of color, design, or look across the entire office to give it a united appearance. If the furniture color or style is different for every individual, an office can look like a secondhand store and give off a disconnected vibe. A consistent furniture scheme creates calmness and synergy.

Balance functionality with form and price. Do the desks have enough storage, are the drawers easy to access, can you stretch your legs and move them freely under the desks or tables? What feels comfortable and functional can depend on several different factors, such as the kind of work employees do, and their height and weight.

Provide your office with the right furniture and remember to leave enough room for your team to be comfortable. Before buying furniture for your office, walk through the rooms and visualize the layouts for outlets, lights, placement of desks, windows, and doors.

Learning the basics of the psychology of color can be an advantage when choosing furniture. Greens and oranges can give your office a creative and energetic atmosphere. Blue gives a perception of coolness and tranquility. However, while it may be tempting, you may not want to buy furniture that is too colorful; pastel and neutral is better than garish.

The type of material will depend on its use. For example, if you're looking for chairs for the executive conference room, leather is usually a stylish choice. But if you're looking for a desk chair for everyday work, mesh is a good option since it provides ventilation for the user. For furniture in high-traffic areas, like your lobby or break room, consider something durable and easy to clean.

BATTERIES AND CHARGERS

Use rechargeable batteries whenever possible. Unplug battery chargers when the batteries are fully charged or the chargers are not in use. A typical battery charger uses around 2 watts when it's plugged in and switched on but doing nothing. Turning off or unplugging one battery charger saves about 10 pounds of CO_2 each year.

To protect against power outages invest in uninterruptible power supplies (UPS) which combine surge protectors with battery packs. These will run a computer for a short time and prevent the loss of information when the power goes out.

Studies have shown that using rechargeable batteries for products like cordless phones and PDAs (personal digital assistants) is more cost effective than using throwaway batteries. If you must use throwaways check with your trash removal company about safe disposal options. Many office supply and hardware stores and supermarkets will dispose of old batteries.

BATTERY BASICS

- Nickel Cadmium (NiCd)—the NiCd is used where long life, high discharge rate, and an economical price are important. Main applications are two-way radios, biomedical equipment, professional video cameras, and power tools. The NiCd contains toxic metals and is environmentally unfriendly.
- Nickel-Metal Hydride (NiMH)—has a higher energy density compared to the NiCd at the expense of reduced cycle life. NiMH contains no toxic metals. Applications include mobile phones and laptop computers.
- Lead-Acid—most economical for larger power applications where weight is of little concern. The lead-acid battery is the preferred choice for hospital equipment, wheelchairs, emergency lighting, and UPS systems.
- Lithium-ion (Li-ion) —the fastest growing battery system. Li-ion is used where high-energy density and lightweight are of prime importance. The technology is fragile and a protection circuit is required to assure safety. Applications include notebook computers and cellular phones.
- Lithium-ion Polymer (Liion polymer)—offers the attributes of the Li-ion in ultra-slim geometry and simplified packaging. Main applications are mobile phones.
- Remember that some of these batteries contain nonsustainable and toxic materials.

Donate older equipment. Many nonprofits, schools, churches, and other charitable organizations are in need of computer and office equipment. Look in your phone book for resources or go online to find national recycling programs. And always separately recycle batteries, considered hazardous waste for good reason (see chapter 16).

3

Employee Engagement
Keeping the Workforce Interested and Enthusiastic

Do you see employees just going through the motions like the walking-wounded carrying discontent, discouragement, and dreariness like a Typhoid Mary, contaminating others and infecting the company culture?

Don't snicker. A Gallup poll estimates that an "actively disengaged" employee costs an organization approximately $3,400 for every $10,000 of salary and those employees combined cost the American economy from to $450 to $550 billion a year due to lost productivity.

According to a 2012 Bersin & Associates report (a talent management, research, and advisory services company now part of Deloitte, a multinational professional services firm), companies spend almost three-quarters of a billion dollars on improving employee engagement. That's a lot of money to find out about employee commitment, and to assess how personal goals and values align with the organization and its culture, contributing to employee happiness and satisfaction.

Disenchanted millennials, many of whom graduated from college with pricey degrees and mountains of debt, are the least-engaged generation. The *New York Daily News* reports that nearly 70 percent of US employees are miserable at work. One theory suggests that "motivation" factors, like recognition, growth, autonomy, and enjoying the work itself are what lead to job satisfaction. But according to a report from Deloitte, 88 percent of employees do not have passion for their work and never contribute their full potential. Their report also found that only around 20 percent of senior management is passionate about what they do, which is an even bigger problem. If you are experiencing this yourself, or seeing this kind of lack of engaged energy at work, read on.

HAPPINESS AND ENGAGEMENT

You might think that when someone is engaged in their job, they're happy. But in the workplace, job contentment and fulfillment are two different things that are intertwined like love and satisfaction—they certainly go well together, but when at polar opposites, the situation can be miserable.

Employees can be reasonably content at work, but if they are not receiving enough recognition, communication, salary, feedback, or opportunities for growth or advancement, they may never be truly engaged in their work. Consequently, they may become bored or disenchanted and might even leave. That is expensive and a waste of energy.

Herzberg's Two-Factor Theory, also known as Motivation-Hygiene Theory, says that motivational factors, like recognition, growth, autonomy, and getting good at and enjoying your work are what lead to job satisfaction and engagement. When managers invest in their employees and help them improve at what they do, they'll be much more likely to be engaged and will find ways to do their jobs better and they will be turned on and tuned in to the company's culture. People need to earn a living, but they also want to think that they are getting paid for a job well done. Behavioral psychologist John Stacey Adams also notes in his Equity Theory that a balance between employee input and fair reward from the employer for their contribution is necessary to ensure an optimal workplace relationship.

The flip side is job dissatisfaction born from factors like poor physical work environment, ineffectual management, arbitrary salary, shaky job security, uninspiring work, and no chance for advancement.

COMPANY CULTURE

A simple definition of a company's culture is basically the company's vision of and for itself, its employees, value system, beliefs, assumptions, habits, and manner of doing business, along with a blend of beliefs, taboos, symbols, rituals, and myths. According to another Deloitte report, the number one concern for CEOs and senior human resources (HR) leaders is "culture and engagement." Creating a good company culture where employees can be their best takes a lot of work.

Job turnover at companies with a rich, internal culture averages about 13.9 percent per year, while those without experience turnover rates as high as 48.4 percent, according to *Entrepreneur* magazine. The rewards are attracting and retaining the best talent, keeping happy and productive employees, and increasing success right down to the bottom line.

PEOPLE FIRST

Rigid, bureaucratic, and rule-bound organizations that were the model of businesses back "in the day" should be extinct. In too many organizations, rules are enforced by intimidation. This couldn't be more counterproductive; it causes employees to disengage both mentally and physically, and can result in absenteeism, ennui, resignation, or termination.

The focus should be on the people in a corporate culture, and should include mapping and guiding the quality of the work life of employees, providing appropriate responsibilities, training, and engaging opportunities. Employees need to feel inspired and valued at work. Scheduling semi-regular one-on-one talks with managers and individual employees where people can address and discuss their concerns directly reassures them that they are appreciated and that their input is valued and meaningful.

Employees also have to feel safe when they take the initiative to try something new. It is management's job to provide people with a safe haven for communication, no matter how difficult the subject. Management should depend on employees to do the right thing on their own instead of tethering them to policies and procedures that force them to do so. Some freedom and self-discipline will empower employees and increase job engagement, resulting in a more efficient work crew and a better bottom line.

Don't be afraid of conflict, it's important to embrace it. The conversations may be tough to handle, but it's part of a leader's job. And the upside of controversy handled well is improved relationships and trust. But be careful, as this can be good or bad, even ugly.

ENGAGING EMPLOYEE ENERGY

One of the foremost forces of any company's resources and power is its employees' constructive energy. Ask any successful person and the first thing you will hear about is passion for what they do. Not everyone is going to be happy every day, but have you heard that old saying "fake it till you make it?" Well, this applies to a positive attitude too.

A Gallup poll found that only half of the people interviewed are clear about what's expected of them at work, and that certainly doesn't engage them. People deliver better results when they have an understanding of what's expected of them. And they're happier when they feel their work is part of a greater picture.

"OUTSIDE ME" VERSUS "INSIDE ME" MOTIVATORS

Outside, or extrinsic motivation, occurs when people are driven to engage in a behavior or an activity to earn a reward or avoid a punishment. For example, a person might do a chore to avoid being disciplined.

Inside, or intrinsic motivation, involves engaging in a behavior because it is a personally rewarding activity for its own sake like training in order to win a contest or being good for goodness sake.

To counter problems of workplace inertia, try the following approaches to create support, engender energy, and persuade employees to take pride in their work and their company.

MEANS OF MOTIVATION

Perquisites or "perks" used to be something regarded as a special right or privilege enjoyed as a result of one's position. It came as a matter of course with some executive status, like access to the company dining room or an expense account. In the past most companies frowned on perks for low-level employees, and only offered them to upper management. Not anymore.

Many companies are finding that perks running the gamut from discounts to vacation spots, sponsored meals, company retreats, job training classes, physical exercise equipment, massages, fresh-squeezed juice stations, yoga classes, and even low-cost loans drive employee engagement, happiness, and loyalty to the company. In the short run they may be expensive, but for the most part they are more than worth the cost.

However, only the right perks will drive dedicated engagement. Friday night drinks will certainly engage drinking buddies. But to drive contentment and/or engagement, the perks have to be in the right place, for the right reasons, and be reasonably significant. Don't confuse a free lunch or a ping pong table with the following substantial perks.

PARENTAL TIME OUT AND PAY

One of the most important benchmarks of a person's life is having a child. It's a wonderful experience but one of the most stressful times of an adult's life. The United States is the only first world nation that doesn't legally require paid maternal or paternal leave benefits. When most companies allow time off, it's without pay. The most popular and forward-paying companies allow both paid maternal and paternal leave—whether paid or not, it's a valued perk, and paid leave is a heavyweight benefit that won't be forgotten.

One more thing—the United States ranks fifty-ninth in voter turnout in the world. Requirements and time off for voting vary state by state. A few states and some companies are requiring or offering time off with pay to vote—a good community perk. Check your local rules.

CLASSY COMPENSATION

The idea is to encourage employees to take instruction on their free time to get better at what they do. Businesses should look at paying all or part of the fees for coursework related to their employees' occupations, and should also encourage employees to take classes they're interested in that have nothing to do with their jobs—to demonstrate that company people are more than just employees.

DONATION-MATCHING PROGRAMS

Research is clear about employees wanting to work for companies that support their communities. More than 85 percent of millennials correlate their purchasing decisions and recommendations to the socially worthy efforts a company creates. Giving employees time off to make contributions is a great way to let them know that their employer supports giving back to society.

FLEXTIME

The ability to plan and employ flexible schedules is one of the most popular employee perks companies can offer. Plus it costs nothing (see chapter 5 regarding alternative job scheduling).

UNRESTRICTED VACATION

Bruce Elliott of the Society for Human Resource Management (SHRM) claims that there is little abuse in companies, both big and small, that use unlimited time off. However, only 1 percent of companies nationwide offer unlimited vacation and according to data from SHRM, only another 2 percent are considering adopting the policy in the near future.

Employees appreciate the trust and flexibility offered by such a policy, and some employers use it for recruiting. With the "honor system" time off, employees can recharge at their own pace; it is a good way to incentivize employees to take the breaks that could bolster their work output and relieve burnout.

Of course there are some hitches too. It's not always fair because not everyone can be out at the same time. Employees might initially take less vacation time and many workaholics won't take advantage of it at all. Plus, people will always consider peer pressure. How many vacation days can you take before your manager or colleagues start to feel you're slacking off? How many days until your coworkers start resenting you? And for some jobs, like first responders and some vital management positions, the policy just isn't feasible.

People can be overwhelmed by unlimited choices. Many employees decide not to take advantage of the policy because it's too hard to figure out or they're afraid about taking the right amount of time off. There is no such thing as a perfect policy for every company. For starters, employees could be allotted a fixed number of days off and be required to take those days off each year. Managers can approve special treatment for unique circumstances, and of course, flexibility should be considered.

COMMUNICATION

The backbone of relationships, personal or professional, is communication. Research from Gallup has shown that managers account for 70 percent of the inconsistency in engagement due to lack of proper communication. Gallup's research has also found that companies more often than not fail to make the right person a manager. Furthermore, research by Clear Company (a human resource management firm) says that, "86 percent of employees and executives cite lack of collaboration or ineffective communication for workplace failures."

It's common for modern-day employees to go through the workweek feeling uncomfortable with the concept of speaking their minds. Management should make employees feel confident when speaking out and let them know that they can do so without fear of criticism, judgment, and reproach. Initiate ways to communicate—be open to anonymous suggestions, face-to-face conversations, and group talks. Beneficial ideas and solutions that could help to push your organization forward are easily lost when people are not at ease discussing them. Listen and involve your team members in decisions that directly affect them.

An open-door policy is essential, and should be encouraged. Institute informal roundtable discussions where people can have a say to improve the quality and to improve the frequency of feedback that employers and employees need. The purpose of giving feedback should be to start a dialogue for shared understanding and to build a bridge for communication. The more you promote an open, safe, and secure space for communication, the more likely it is that your staff will bring great ideas and constructive dialogue to the table.

Start by recognizing specific behaviors you want to praise, mention positive results or how some action impacted other employees, the company, clients, or community. Offer gratitude and encourage more of the same. But be careful as an overabundance of praise can feel insincere and if done too often, it loses its significance and influence.

Don't dwell on the past. Feedback about unmet expectations is necessary, but it should be forward-looking; don't rehash old examples—be focused and positive.

Avoid choosing favorite employees and spread praise around. Leave out the word "but" and don't sandwich negative feedback between positive feedback. As a result, you will be able to create clear, shared meaning and motivate people to action.

Management should engage and motivate leaders to know what their people need and want. Employees have be informed—through transparent communication with their coworkers, managers, and customers—about what's going on in the organization, and about how they can help and maintain their place within it. Workers will be far more inclined to support what they help to create.

Every time a manager is in front of an employee, whether one-on-one or in a group, there is an opportunity to increase that engagement through open-ended dialogue. Unlike questions that give people limited options for response, open-ended questions encourage people to express their opinions and ideas.

Respect is not only demonstrated by listening and collaborating but by getting the right people in the right room at the right time to confront a challenge, whether individually or in a group. Leaders need to create a more inclusive environment for everybody, regardless of role, experience, age, or any other factor. Ask questions such as, "Are there any projects you'd really like to work on?" A leader has the responsibility to find out the sources of employee frustration. Ask your employees if everything is explicit; ask them what they think might be positive changes to make; find out what they like most or least about the company.

COMMITMENT TO PARTICIPATION

Create a comprehensive employee syllabus or manual that is clearly and simply written and that includes procedures for handling every imaginable scenario. Ask employees for their ideas for the manual so they feel a sense of ownership.

Host special employee events where families can be involved, such as picnics, fairs, workshops, company or employee milestones, or host a day at the office. The more sense of "family" you can create the more productive and engaged employees will be with the company and their fellow workers.

Finally, have designated charities where people can donate both money and time. This helps each person feel that they are involved in community-business participation. Research indicates that people feel better and have enhanced lives when they volunteer or contribute. It also helps the company's bottom line by increasing employee performance, engagement, and goodwill by demonstrating that their company cares about something outside of business.

BE REASONABLE ABOUT EXPECTATIONS

Don't ask for too much too often. To avoid burnout and discouragement, set short-term goals leading to larger projects. Work is like exercise. Don't expect six-pack abs after a week of crunches. Results can take time and people need to be urged to push their limits.

The minimum that employees should expect from employers is that they will pay them on time and provide some benefits, as well as fair treatment, a safe working environment, access to tools and materials needed in the scope of their jobs, and reasonable aids to become fully engaged in their work. They have the right to expect that their employers will treat them respectfully and adhere to anti-discrimination laws that prohibit unfair employment practices based on non-job-related factors, such as race, national origin, religion, and gender. Additionally, workers should expect management to provide them with unambiguous communication and clear direction as to their job duties, rewarding good performance accordingly, as well as appropriate recognition and feedback.

Employers have the right to expect employee loyalty to management and to the company. Honesty and trustworthiness in dealing with the company involves telling the truth in all work-related matters. Workers should be dependable and reliable, showing up on time and carrying out the tasks and duties required in the scope of their employment. They need to be willing to see that delegated work gets done and be diligent about fulfilling their responsibilities. Self-discipline and self-responsibility involve accepting accountability for your own actions and using your talents and abilities for the betterment of yourself and the company. This also requires that you learn to handle your emotions.

Finally, challenge employees to work outside their comfort zones—even outside their job descriptions. Boredom is just another path to disengagement and unhappiness.

As David Letterman might say, "The top ten reasons for employee contentment are. . ."

1. Appreciation for good work.
2. Good relationships with superiors.
3. Good relationships with colleagues.
4. Pleasant work environment.
5. Company's financial stability.
6. Learning and career development.
7. Job security.
8. Attractive salary.
9. Interesting job content and opportunity.
10. Shared company values.

And the top ten reasons for discontentment are:

1. Limited career growth.
2. Micromanagement frustration.
3. Lack of progress in job or company.
4. Job insecurity.
5. Unsupportive boss and little or no confidence in company leadership.
6. Poor communication.
7. Unpleasant coworkers.
8. Lack of meaningful work and boredom.
9. Overworked and underpaid.
10. Work and life imbalance.

What employees desire and require:

- Value and respect.
- Communication.
- Advancement and reinforcement.
- A long-term investment in growth from leaders to ensure that their skills stay ahead of the curve.

- Confident leaders who are not threatened by employee potential.
- Clear purpose and genuine responsibilities.
- Autonomy, freedom, and trust.
- Flexibility.
- Feedback—the more positive, the better.
- Transparency.
- Open mindedness.
- Fair compensation.

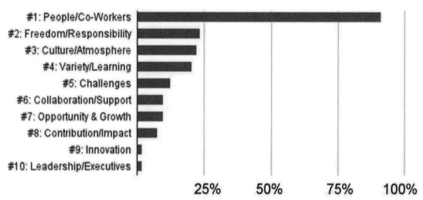

Figure 3.1.   What Do You Like Most About Work? Source: https://www.dol.gov/.

EMPLOYEES NEED TO FEEL VALUED

Employees have to be recognized. This covers both the quality and the frequency of recognition that employers and employees receive and give to each other. It's about showing employees the respect they deserve and letting them know that the company notices, values, and appreciates all their hard work. In fact, many social scientists claim that if more money is the only reward for a task, it can actually lower employee motivation. It's called the "overjustification effect" and results when an external incentive decreases a person's intrinsic motivation to perform an action.

One top reason that people leave their jobs is that they don't feel appreciated. A survey conducted by Deloitte found that, "Organizations with recognition programs that are highly effective at enabling employee engagement had 31 percent lower voluntary turnover than organizations with ineffective recognition programs." Moreover, 90 percent of employees stated that recognition increases their bond with the company, their dedication, and their initiative in their work, and reduces frustration.

A 2012 SHRM survey found that, "Companies with strategic recognition reported an employee turnover rate that is 23.4 percent lower than companies without any recognition program." They also discovered 41 percent of companies that use peer-to-peer recognition have marked positive increases in customer satisfaction, and peer-to-peer evaluation is 35 percent more likely to have a positive impact on financial results than manager-only recognition. The reverse can also be true; ineffectual employee recognition programs can have a major, negative impact on company morale, mutual respect, and engagement.

In addition, scientists know that positive recognition produces the dopamine effect—a pleasant euphoria produced by brain chemistry. Unfortunately it wears off quickly. So the trick is to make a habit of praising employees and they will respond by trying to do even better, both for the satisfaction of recognition and the dopamine high.

Despite years of research proving the overwhelmingly positive effect of employee recognition to the bottom line, few bosses take the time to recognize and reward their employees for a job well done—and even fewer employees report that they receive either recognition or rewards at work. The most effective forms of employee recognition cost little or no money, such as a verbal or written "thank-you" for employees who do a good job, publicly celebrating team and group successes, and one of the best rewards is as simple as a free pizza, according to an experiment carried out by Duke University psychology professor Dan Ariely.

Should something go wrong or if someone makes a mistake, don't "punish" the person or team. Rather, talk it out and teach the correct procedures, and offer support and further instruction when needed. Remember that reprimands will only make things worse and the employee may become angry and bitter and may want to get back at the company. If errors continue after correction, then you may need to evaluate that person to make sure he or she is a good fit for the job.

Employers have an excellent opportunity to make a difference in their employees' lives. This may mean a simple smile, enquiring about their family, or asking about their interests or problems. If you sense that someone is depressed or stressed, help that person get the necessary resources as employees with troubles have higher absenteeism, increased health problems, and decreased performance. Remember that we are all human; we should be working together—to get the best results, we need to care about each other.

The best leaders know that the only way to get things done and move a business forward is through the efforts of dedicated employees. That means leaders need to go beyond merely lifting up those who need extra motivation. They need to recognize the people who exhibit the behaviors they want others to emulate.

ANNUAL PERFORMANCE REVIEWS DON'T WORK

Most managers hate annual performance reviews, employees dread them, and many companies are getting rid of them. Instead they are opting to exchange reviews on a more short-term basis and using alternative techniques like one-to-one or group meetings, making sure they are recurrent, and that everyone can attend.

After each meeting, review conversations and agreements with an e-mail to individuals or groups. Make very specific recommendations (with examples) on what can be done to improve performance. Make sure employees understand that any criticism is about the work and not the individual. Try not to be negative—instead of telling someone that their report stinks, suggest that you get together for an editing session.

HELP YOUR EMPLOYEES GET BETTER ACQUAINTED

In big organizations, employees can get through the day without knowing who's on the same floor. The larger the company, the easier it is to get lost and to be ignored. It's important to encourage employees to get to know each other, not only to encourage them to share resources and referrals but to foster a feeling of belonging, which plays a powerful role in human behavior.

Try a happy hour at the end of the workweek. Alcohol or even coffee cake and ice cream can be social lubricants that may loosen people up and encourage them to communicate with each other on a more personal level without the fear that big brother is scrutinizing. Business talk is important, but a more congenial approach releases tension and breaks the ice of the workplace.

FOCUS ON COLLABORATION AND WORKING TOGETHER

Support interpersonal and interdepartmental communication between people and managers. Collaboration is one of the most important things for businesses to focus on and improve, as it can help to germinate and enhance ideas, cut down on wasted time, and develop and advance employee engagement. No one should be afraid to be vulnerable. When you feel that you are in over your head, appreciate that you are in the process of experiencing growth and that other people are there to offer assistance. Lean on your team members; it feels good for both parties.

One of the best ways to improve collaboration in your company is to implement social intranet software (a local or restricted communications network). This serves as a virtual platform for employees to communicate, especially when they feel shy or disconnected. Once people feel comfortable with it, social intranet involvement can turn into an incredibly powerful tool, serving as an important method of connecting employees and expanding information resources and the business.

To build a truly amazing team, find people with complementary skills. If people all think the same way, it inhibits diversity. But remember, complementary skills also come with different personalities, communication techniques, and interests.

Employees should not waste time being jealous or carrying grudges. Confucius says, "Before you embark on a journey of revenge, dig two graves." Practice "extreme resilience"—that is, the ability to recover fast from adversity, spending little time on self-flagellation and adversity. Move on.

Allowing staff to move from one department to another will help employees get to know each other better and begin to see the ways in which other departments operate, thus leading to a more engaged, considerate, and supportive workforce.

Colleagues, including managers, who know each other well and who communicate often, keeping interactions as positive as possible, help to create a supportive environment and better the company for everyone.

SERVE AS A MENTOR AND AS A GOOD EXAMPLE

Like any other leader, such as a parent, teacher, or boss, an employer needs to learn when to let go. Employees need to feel valued and challenged and trusted with some freedom to explore and learn within the job. A good boss sets clear expectations, directions, and boundaries, and provides beneficial leadership and sound direction. And also knows when to unleash. Don't forget the personal touch by talking with employees on a first-name basis in group and one-on-one settings.

Be a leader who shares, sacrifices, and communicates. Be confident enough to own your mistakes. An engaged manager doesn't use either real or perceived threats of penalty, loss of privilege, loss of job, etc. Disengaged managers will infect their people with their apathy, negativity, and beliefs that the work doesn't matter.

Employees listen and pay attention to what is said through the grapevine and what has been done around the workplace. Think about what employees are seeing and hearing, and develop the awareness to act the role of the leader you want to be, modeling the actions and characteristics that you would like to see in others. Employees hear what they want to hear and see what they want to see. Start meetings highlighting good accomplishments. When people are praised, it generally makes them want to do better. End the meeting by reminding the employee(s) that they are valued.

WHAT WOULD YOU CHANGE ABOUT THE BOSS?

For *Entrepreneur's New Year Employee Report* of 2015, participants were asked what one thing they'd like to change about their managers. The top five answers were:

- 7 percent would wish for better team leaders
- 8 percent would increase wages
- 10 percent would seek to improve empathy and people skills
- 11 percent would want their boss to quit or retire
- 15 percent would improve communication

And if they were the boss, the top five were:

- 9 percent would modify working hours
- 10 percent would improve wages and benefits
- 11 percent would seek to establish standards for behaviors and company policies
- 11 percent would want to improve communication
- 16 percent would fire, demote, or make other changes to improve employee caliber

WATCH YOUR BODY LANGUAGE

Negative body language can bust you as easily as your choice of words. Excellent nonverbal communication skills tell your audience that you're confident, energetic, engaged, and honest, says Tonya Reiman, author of *The Power of Body Language*. She advises you to stand tall, with your neck elongated, ears and shoulders aligned, chest slightly protruding, and legs slightly apart, distributing your weight evenly. Don't fidget. Show interest in what others are saying by head nods, raised eyebrows, and low utterances, and by slightly mimicking the actions of others. Holding eye contact has tremendous power to persuade. Smile, do not roll your eyes, and don't cross arms over your chest. Use a firm but not crushing handshake—it's not a contest.

ENGAGEMENT MEANS SAVING MONEY

Companies that have high levels of employee engagement received a 37 percent Net Promoter Score (NPS) versus a 10 percent NPS for teams with lower engagement. NPS is a management tool used to gauge the loyalty of a firm's customer relationships as an alternative to traditional customer satisfaction surveys. In other words, when employees are satisfied, they offer better service, find enjoyment in what they are doing, and might even finds ways to do their jobs better. This increases repeat business, brand loyalty, profit, and, voila, more employee engagement.

One the other hand, dissatisfaction and job turnover are expensive. According to the Society for Human Resource Management (SHMR), studies predict that every time a business replaces a salaried employee, it costs about 20 percent or six to nine months' salary on average. But others predict the cost is even more and that losing a salaried employee can cost as much as two times their annual salary, especially for a high-earner or executive level employee.

Research from the Corporate Leadership Council shows that engaged employees are 87 percent less likely to leave the organization than the disengaged. In addition, engaged employees are seven times less likely to have a lost-time safety incident. This is very serious stuff, not only because it's costly, but because it can hamper or ruin company and worker attitude.

Shrinkage is a cute business term for a company getting ripped off by its employees, to the tune of billions of dollars each year. Along with theft comes a self-centered attitude that can become infectious.

Absenteeism sometimes can be the result of illness and apathy. It costs roughly $3,600 per year for each hourly worker and $2,650 each year for salaried employees. The links between disengagement, stress, and poor health are all part of this increased absenteeism.

Apathy is closely related to absenteeism, carelessness, boredom, and theft. Gallup research shows that employees who have quality engagement in their work have 41 percent fewer accidents. Disengaged employees don't care as much about their work nor do they pay close attention to their jobs, so they're more prone to making mistakes, to the tune of 35 percent more work-related accidents and a half-trillion dollars annually, according to a recent Gallup poll, and are counterproductive to their company.

In a report from Development Dimensions International (a human resources consulting firm), an unnamed Fortune 100 manufacturing company reduced quality control errors from 5,658 parts per million to 52 parts per million by focusing on engagement. Sometimes referred to as the "working dead," disengaged employees have a major effect both on clients and on the bottom line of the US economy.

4

Workplace Wellness
Keeping the Help Happy and Healthy

You don't have to be "Silicon Valley" rich to offer wellness at work via fancy exercise machines, personal trainers, chefs or masseurs on command, or by building a gymnasium to encourage a healthy workforce. Wellness programs are possible no matter how big or small the company or budget, but first it has to be made a priority. There are lots of simple and inexpensive alternatives to keep employees healthy and happy. Even small businesses can benefit their employees by providing on-site resources such as comfortable chairs, stand-up desks, and minor medical screenings.

According to the *Wall Street Journal* and the iOpener Institute, happier workers help their colleagues a third more often than unhappy ones, achieve their goals 31 percent more often, and are 36 percent more motivated in their work. This should be of great interest to employers as more companies recognize the benefits of satisfied and engaged employees who enjoy good moods and good health.

FACTS AND FIGURES

- More than 1 million people miss work every day due to stress at work, costing $225.8 billion a year in the United States.
- Seven in ten American adults are overweight or obese, and 29.1 million Americans (9.3 percent of the population) suffer from diabetes. To avoid developing Type 2 diabetes, keep an eye on diet—eat more plants, cut down on animal fat, drinking, and smoking, and eat nutritious nonprocessed food.

- More than 80 percent of employees feel as if companies are expecting too much from them and their associates and claim that their company doesn't do enough to promote employee health.
- Eighty percent of US workers are putting in 48-hour workweeks. One-quarter of the workforce labors six or more hours a week without getting paid for it, and 47 percent of management does the same, but leisure time has decreased by up to 37 percent.
- Seventy percent of employees say they have to work overtime in order to stay ahead and 62 percent of workers have reported that their workload has caused them to put off vacation time.
- Sixty-nine percent of employees say that their main cause of stress is work, 41 percent say they feel tense or stressed out throughout the day.
- Thirty percent of working mothers would take a pay cut to spend more time with their children and 25 percent of the workforce has missed a significant family event due to work, causing family conflict.
- Depression results in more days of disability leave than other chronic health problems.

On the other hand, wellness programs ranging from simply offering information to workers to subsidized healthy lunches, fitness education, emotional counseling, a company gym, pet-friendly environments, and other sympathetic employee plans realized:

- Fewer absences per employee and 25 percent lower sick leaves, workers' compensation, and disability insurance costs.
- Eleven percent more revenue per employee and 28 percent higher shareholder returns.
- In one company, 57 percent of people with high health risk reached low-risk status by completing a worksite cardiac rehabilitation and exercise program.
- In 2001, MD Anderson Cancer Center created an employee health and well-being department, staffed by a physician and a nurse case manager. Within six years, lost work days declined by 80 percent and modified-duty days by 64 percent. Cost savings calculated by multiplying the reduction in lost work days by average pay rates totaled $1.5 million, and workers' compensation insurance premiums declined by 50 percent.
- The giant pharmaceutical company Johnson & Johnson estimated that their wellness programs have cumulatively saved the company $250 million on health care costs from 2002 to 2008, or a return of $2.71 for every dollar spent.

ENCOURAGE BEING WELL

A healthy workforce results in reduced downtime due to illness, improved morale, increased productivity, and higher employee retention, while employees get the benefit of increased job satisfaction and an improved ability to handle health and stress.

Something as simple as stocking healthy snacks at work can improve levels of engagement in your organization. Organize social events and encourage behavioral changes at all levels of the organization. Consider developing a Web site customized to your wellness program. Profile employees who are making changes and embed health advocates to encourage staff.

Physical and emotional well-being is important for everyone, yet everyday stress and other factors cause people to neglect healthy habits and can encourage bad ones that are like a bed—easy to get in to, hard to get out of. Eighty-eight percent of employees polled believe it's crucial to have a healthy work-life balance, as well as a positive atmosphere in the workplace. In addition, in a study from Jobsite states 70 percent of polled employees say that cultivating friendships at work generates a positive influence on their productivity and happiness.

Take a look at any employee whose engagement level seems to be suffering, and you might find an employee in distress. Employee wellness (energy levels, interpersonal relationships, sleeping and eating habits, meaningful communication) has a direct affect on productivity, so it is important for employers to keep an eye on this. Regular lunches or roundtable discussions with coworkers encourages people to talk about health and happiness at work and at home.

A happy employee is also a healthier one, which means lower healthcare costs. A Gallup poll found that the top 25 percent of engaged workers had 50 percent fewer accidents, fewer lost days, and a reduced number of visits to the doctor.

And by the way, satisfied and engaged employees perform an average of 20 percent better than dissatisfied ones, they're 87 percent less likely to change companies, they outperform their competitors by as much as 20 percent, and are they're three times more creative than their disengaged counterparts.

A major cause of mistakes is simple fatigue. The brain gets tired when focusing on one thing for too long. An employee should only work on one task for a set period of time, as they lose concentration for multiple and complex tasks the more difficult and time-consuming they become.

The National Institute for Occupational Safety and Health (NIOSH) has developed a suggested list of leading work-related diseases and injuries. Causes include:

- Boring, repetitive, machine-paced work tasks, and role vagueness and uncertainty.
- Lack of control over one's work.
- Nonsupportive supervisors or coworkers.
- Limited job opportunities.
- Rotating shift work.
- Poor physical environment and social support.
- Heavily induced stress.

People appreciate an inquisitive or sympathetic ear. And it wouldn't hurt to have a counselor in your rolodex, or even a confidential number of a friend or family member to call when someone might need a sympathetic ear. Look out for:

- More than average number of sick days.
- Anxiety, irritability.
- Behavioral problems.
- Substance abuse, sleep difficulties.
- Sleep complaints, headache, and gastrointestinal symptoms.
- Inferior work and productivity.
- Uncommunicative conduct.

BEING WELL

When people think about wellness they generally consider biking to work, gym memberships, stretching or walking, and healthy food in the cafeteria or from food dispensers. But also consider drastically disgruntled employees, sometimes characterized in the past as "Going Postal," who harm themselves or others, either physically or mentally. Remember that mental as well as physical health issues in the workplace are of the utmost importance.

Changes for better health can start small. A study done by researchers at Virginia Commonwealth University found that a dog-friendly workplace boosted morale and reduced stress levels, whether people had access to their own pets or those belonging to others. This doesn't cost anything, and shows employees that management can create a warm and family-like environment.

Many health workers believe that employers have a responsibility to provide employees with the methods they need to live healthier lives—and to provide the time during the workday to pursue those methods. If employees are suffering from chronic conditions and leading unhealthy lives, it will impair their ability to do their jobs. Especially during the businesses' busiest times at work, it's critical to eat right, exercise, get adequate rest, and take steps to manage stress.

APPROACH A WELLNESS PROVIDER

Speak to your insurance company, a community mental health consultant, or organizations that deal with the issues of on-the-job physical and mental health to discover what resources and incentives are available to your employees. Unfortunately, there is still some stigma attached to seeking counseling for psychological stresses. Many companies have an employee trained to provide first aid but rarely is anyone on staff to address concerns regarding mental health, which is the most prevalent cause of illness among people of working age. Appointing someone as a mental health mentor or training a number of personnel in psychological awareness would make a huge difference in employee awareness and give people the ability to seek help when needed.

Evaluate your workforce's overall health and conduct a survey of employees to gauge their interest in a comprehensive wellness program. Speak to your insurance company or health consultant about evaluating the company's claims history, chronic conditions (but be aware of privacy issues), and how the company can influence areas that can be affected by lifestyle changes.

Ask employees or form a committee to discuss what techniques they would prefer, which will also encourage team building. Keep in mind that results will take time and dedication, as they do for any individual making lifestyle changes.

CREATE AN AMIABLE AMBIANCE

The Chinese call it feng shui (pronounced fung shway), the ancient Chinese art of placing objects that will bring either favorable or unfavorable effects in a living and/or working space. Many businesses in this country that have embraced this philosophy have enjoyed benefits when learning to use the flow of energy called "chi" to create a peaceful, pleasant, productive, and prosperous workplace. In many ways it is similar to the concept of ecotherapy— part of an overall philosophy of our connection to the natural world and the environment in which we work and live. Remember that dissatisfaction at the workplace comes from factors like the physical and emotional "hygiene" of a work environment. Getting your office environment "healthy" will pay off with a multitude of health benefits.

It's a bummer to look forward to drudgery in a drab, dimly lit, and cramped cubicle. An office should be pleasing to work and live in. Look into employing an office design facilitator who can help make the office amiable—this will help enhance work output and create a pleasant and comfortable ambiance in the workplace. Paying attention to the office environment will make coming to work less stressful, and even enjoyable.

At work, we all need a home away from home. Allow employees to personalize their work space within reason, especially if two or more workers share an area. The interior configuration of workspaces and resting areas are factors that affect concentration, confidence, and creativity. At times productivity becomes impossible if there are noises that create unwanted distractions causing frustration and a decline in employee productivity and morale. If absolute quiet is necessary, provide areas for concentration with noise-canceling headphones and the use of soundproofing.

Make sure all equipment is designed ergonomically to provide optimum comfort, to avoid stress or injury, and to physically motivate workers by giving them the right tools to carry out their tasks. For example, use a laptop or screen support to make sure that the screen is positioned correctly for the employee so they adopt the best sitting posture, and use a palm rest for the mouse to ensure that the hand, wrist, and forearm are properly aligned. Use footrests when the feet cannot be rested flat on the floor naturally.

Plants and natural light can enhance the beauty of your surroundings and clean pollutants out of the air as they add oxygen and humidity to the indoor environment. Bud vases and a bouquet or two of flowers can brighten up a workplace making it pleasant and more connected to nature. Even if plants are not an option, pictures or murals of outdoor scenes offer some relief.

Offer a bowl of fruit in the cafeteria or break room. Healthy food helps people work and think better, improves mood, and increases energy levels.

Watch for tricky detrimental energies from electromagnetic fields emitted from high tension wires, industrial radar, microwave beams, smart meters, and other electrical appliances that may be dangerous to mental and physical health.

Scientific studies confirm that colors bring about emotional reactions in individuals and are useful in stimulating creativity, intensifying the intellect, and heightening motivation. For example, warm colors, such as orange, red, and yellow can cause people to think the temperature in the room is warmer than it actually is and cool colors, such as blue, green, and light purple cause people to think the temperature is cooler. Research has linked green with broader thinking and more creative thought, and red may reduce analytical thought. Blue is a favored color and will satisfy the majority of people while yellow has the opposite effect. Orange is related to good value, pink is calming, and white can lead to boredom.

Music can affect emotional well-being, physical health, social functioning, improved communication abilities, and cognitive skills.

Aromatherapy aids psychological and physical well-being. The EPA informs us that six out of ten buildings are environmentally "sick" and that indoor air quality is worse than outdoor quality, which is the United States' number one cause of respiratory health problems.

MOM AND DAD IN THE WORKPLACE

Make the workplace family-friendly. Balance of life and work is a major stress factor for people. Therefore, allow workers to take time off for school events or to stay home with ill children without using sick or vacation days. If possible, offer childcare near or on premises. Research has shown when employers subsidize childcare, it saves money from absenteeism and creates more stability and confidence at work. Thirteen weeks of (unpaid) maternity leave is mandated for parents and a 2008 report from the Families and Work Institute indicated that 16 percent of companies with at least 100 employees provide full pay during maternity leave. This is down from 27 percent in 1998, and just 14 percent of US employers offer paid leave for new dads, according to a recent study by the Families and Work Institute.

Nearly half of Americans think companies should be obligated to provide paid time off to new fathers, according to new HuffPost/YouGov poll. However, 36 percent of those surveyed said companies shouldn't be mandated to give new dads paternity leave, while 15 percent were unsure about the issue.

BREAKING BAD HABITS

Focus on small initiatives that could offer big influences, like offering free nicotine patches to workers who want to quit smoking, using reminders on computers to prompt stretching during the day, and starting a weight loss program.

GETTING TO KNOW YOU: CELEBRATE STAFF BIRTHDAYS, MILESTONES, AND ACCOMPLISHMENTS

Isolation can be a major stressor. When employees feel they are part of a team, they demonstrate improved communication and social interaction, and can even achieve more with less direction. A few simple ways to encourage this is to celebrate birthdays and special occasions, start get-to-know-you meetings that are not about work, arrange company outings, happy hours, or potlucks to encourage mingling during lunch breaks or at casual parties.

When employees feel like they're getting the proper amount of attention, they will contribute more and be further engaged in their work. Employees' lives are full of big and small but important moments, both inside and outside of work. Throwing celebrations will give your employees recognition for accomplishments and highlights in both their business and their personal lives.

Employee morale will get a boost when staff members come together for a singular purpose. This helps to create a sense of team unity for the employees, especially if you are celebrating a team accomplishment, and will motivate people as well. Seeing others receive recognition or individual appreciation encourages an employee to continue to work hard.

However, avoid creating mandatory social engagements. The best way to encourage involvement is to choose outings and team-building activities that fit with your company's culture. You could even double-dip the good feelings and provide some charity work with your team-building activity.

If you are stumped about what sort of philanthropy would engage your team, simply ask them to offer suggestions on what charities have a special place in their hearts. Implement social interaction at your next meeting by asking everyone to start by sharing a little-known fact about their hobbies, avocations, or special interests.

And how about something for everyone, like an Employee Appreciation Day that comes once a year to show that they are appreciated all year?

FUNCTION ON A FIRST-NAME BASIS

Getting to know each and every staff member on a personal level can be quite a challenge, especially with a large staff, and learning everyone's name by heart can seem next to impossible. But it's important to try. Operating on a first-name basis can create a number of benefits.

Everyone needs to feel recognized as more than just another cog in the machine. When an employee is called by his or her first name, they feel acknowledged. Even if you slip up from time to time, your staff will appreciate the fact that you're putting in the effort to get to know them on a more personal level. Try gimmicks like mnemonics—even name tags for a while, to help the process.

PERSPIRE TOGETHER, ASPIRE TOGETHER

With the rate of obesity skyrocketing, helping employees shed extra weight should be a priority. Make losing weight a philanthropic event by offering money to a favorite charity for pounds lost. For others with more serious problems, have an insurance agent, consultant, or a referral service on hand.

Obesity increases the risk of chronic health problems like heart disease, Type 2 diabetes, and some types of cancer. Make arrangements with a local gym to give employees a discount on membership. Have a trainer from a gym come in and talk about the role exercise plays in weight control and overall health.

Exercise is one of the best ways to relieve stress. Sponsor a 5K walk/run on a weekend and give prizes to employees who complete the route in the best times.

Encourage employees to train for the event ahead of time. Give everyone something for participating, and end the event with a picnic or party. Encourage people to take a short walk when they get blocked or stressed, especially after eating. If it's affordable, introduce exercise machines or treadmill desks to allow employees to work at their computers and exercise at the same time.

Some companies pick up all or part of employee fees for marathons, ironman, or other physical contests and activities, as well as diverse fitness and wellness programs. Make room for bikes to be stored safely to discourage theft.

Some companies play exercise videos in the office and schedule times when employees can exercise together. It's not only a healthy stress reducer, but also encourages conviviality.

Arrange for a local nutritionist to come in to give a series of talks on how to cook and eat better. Invite employees to share their own healthy recipes and include tips and ideas for staying active in a company newsletter.

Offer pedometers to wear to record the number of steps people walk each week. Hold contests to see who can log the most steps weekly.

Recently, researchers at Stanford University tested creativity in people who were walking versus sitting and found that creativity improved by an average of 60 percent when the person was walking.

SIDESTEP SHARING SICKNESS

Advise all employees to stay home when sick and for 24 hours after they no longer have signs of a fever (100 degrees Fahrenheit or 38 degrees Celsius) even if they feel better. Make sure a fever is gone without the use of fever-reducing medicines containing ibuprofen or acetaminophen that may mask or suppress symptoms. Employees who get sick at work should go home as soon as possible. If a person cannot go home immediately, he or she should be separated from other employees.

Encourage time off for all employees who want to get vaccinated. Provide resources and a work environment that promotes handwashing, provides tissues for covering coughs and sneezes, and cleans surfaces and items that are likely to have frequent hand contact.

DECOMPRESS STRESS

Anxiety disorders affect 40 million adults and are the most common psychiatric illnesses in the United States.

Stress can aid in inducing almost any disease from skin rashes to strokes, constant fatigue and chest pains, to insomnia and mental illness. Stress not only makes it more challenging for employees to do their jobs, it also reduces resistance to illness—and that means more days missed from work.

When stress goes up, productivity goes down. People shouldn't push themselves to be perfect or to do tasks that can be done tomorrow. Avoid being overly critical about yourself or others.

The best stress shrinker is talking. Sharing thoughts and feelings with another person, both supportive and empathetic, whether at home or at the workplace is essential. If you don't feel that you have anyone to turn to, it may be time to talk to the human resources department, a doctor, or a therapist at a community clinic. A company's health insurance should include mental health counseling.

Try to broaden your horizons by meeting new people, taking a class, joining a club, sharing common interests, or by volunteering your time. Being helpful to others can be rewarding in many ways, offer a new perspective on life, and bring about a feeling of personal satisfaction.

Scientists have found that regular participation in aerobic exercise can decrease overall levels of tension and stress, elevate and stabilize mood, improve sleep, and improve self-esteem. Psychologists suggest that a 10-minute walk may be just as good as a 45-minute workout. Exercise can work quickly to elevate depressed moods. In one study, researchers found that those who got regular vigorous exercise were 25 percent less likely to develop depression or an anxiety disorder over the next five years.

- Walk, bike, swim, jog, get on a stair stepper, or dance three to five times a week for 30 minutes.
- Set small goals and aim for daily consistency rather than perfect workouts. It's better to walk every day for 15 to 20 minutes than to wait until the weekend for a three-hour fitness marathon.
- Find forms of exercise that are fun or enjoyable.
- Enjoy audio books, podcasts, or music while exercising.
- Recruit an "exercise buddy" to make it easier for both to stick to an exercise routine.

- Be patient. It takes about four to eight weeks to feel coordinated and sufficiently in shape so that exercise feels easier.

Set up a library with books and audiotapes on stress reduction techniques. Look into a massage therapist who can come in once a week to give 10-minute chair massages. Offer flextime, work-at-home, and job-sharing programs for employees who are dealing with stressful issues.

Depending on expense, interest, and sustainability there are many methods, classes, and activities that a company might consider—from health seminars to Tai Chi, Qigong, Pilates, and yoga, healthy cooking, and meditation to many forms of exercise. Talk to one another and see what people suggest.

Offer health checks to help employees monitor their cholesterol, blood pressure, and blood sugar levels, and provide hearing and vision screenings. This helps employees be more proactive with their health and identifies health problems early.

THE SILENT KILLER: "SITTING DEATH SYNDROME" (SDS)

Most of us sigh with relief when we sit down, and grunt when we have to get up and move. Combine that with time spent driving, watching TV, and other sedentary activities, and we may be logging more hours sitting than in any other position. Watch out, couch potatoes—new science claims that, among other nasty things, sitting too much can kill you!

Too much time on the tushy can lead to a long list of musculoskeletal miseries from sore backs and tension headaches to carpal tunnel syndrome and can strike those whose who don't pay attention to their daily body positions. In fact, scientists claim that the more hours spent sitting, the shorter the lifespan. One study found that reducing the average time spent sitting to less than three hours a day could increase life expectancy by two years.

Sitting for 8 to 10 hours a day (without any exercise) is linked to a wide range of debilitating ailments, including some that are life threatening to more than 300,000 Americans annually. If it was classified as a bona fide disease, it would make SDS the third leading cause of death in the United States, right after heart disease and cancer.

The New England Journal of Medicine found that being sedentary and being out of shape may be more hazardous than other well-known risk factors, such as smoking, hypertension, and heart disease, and can be a contributing cause to no less than two dozen disorders. What's really scary is that it affects nearly three out of four adults and a growing number of children, and is projected to cost the United States $1.5 trillion over the next decade.

One study found sitting for prolonged periods raised the risk of cardiovascular disease by 14 percent, cancer by 13 percent, and diabetes by a whopping 91 percent. Those who sat for long stretches and got no regular exercise had a 40 percent higher risk of early death, according the National Health Service. Moreover, 25 percent of Americans have inactive lifestyles and 75 percent do not meet the weekly exercise recommendations of 150 minutes of moderate activity each week, and muscle-strengthening activity twice a week to maintain good health.

Just being moderately active can cut your risk of coronary disease by 30 to 50 percent. One half-hour of moderate activity most days of the week can significantly reduce the risk of SDS, and in a time-crunched world, just 15 minutes of exercise three times a day is enough to substantially improve health. Taking short walks, climbing the stairs, even playing actively with your kids and pets can add up to real health benefits. But you have to stick with it.

SIT SMARTER

In order to allow the bones in your spine to stack well and permit the muscles alongside them to relax, sit with your behind sticking out behind you, but not exaggeratedly so. Now, when you breathe, each in-and-out breath will automatically lengthen and settle your spine.

This gentle movement stimulates circulation and allows natural healing to go on even while you sit. While conventional advice tells you to tuck in your pelvis to maintain an S-shaped spine, a J-spine—meaning a posture where the back is straight, the lumbar relatively flat, and the buttocks protrude slightly—is far more natural. This biomechanically correct posture allows you to move freely, discourages pain, and allows your digestive organs to function without restrictions or blockages.

By tucking your pelvis, you lose about a third of the volume in your pelvic cavity, which squashes your internal organs and compromises them in a variety of ways. This is further compounded if you're both "tucked" and "hunched" while sitting.

Another way to elongate your spine is by using your backrest as a traction device. Use either a towel or a specially designed traction cushion for this purpose. This simple maneuver brings your back away from the backrest, lengthens your spine, and then roots you higher up against the backrest.

This position helps get traction on your discs, allowing them to rehydrate and prevent the nerves from being impinged between your vertebrae. It will also help flatten out your lumbar area and this alone can sometimes provide immediate pain relief if you have sciatic nerve discomfort.

Employees who work in front of a monitor all day are predisposed to back and neck pain. Help them make the changes they need to prevent aches and pains, perhaps offer an ergonomics course. Whatever, however, and for what length of time is a personal choice, but it's vital to well-being to get up and move, and as Mom says, sit up straight!

AN IDLE BODY AND TYPE 2 DIABETES

It's second nature to say, let's sit down and talk about it, rather than, let's take a walk and discuss things. Like poison, the danger is in the dose. Sitting makes it easy to eat, jogging makes it harder. Workers can burn up to 144 calories per day just by standing for three hours. Sitting also takes a load off your skeleton and muscles and, over time, this may weaken them. When not used after a meal, sugar doesn't get moved in the blood as it should, which can put you at a higher risk for high blood glucose levels. The beta cells in the pancreas respond by producing increased amounts of insulin, a hormone that helps the body's cells absorb glucose to use for energy. Cells in idle muscles don't react as readily to insulin so the pancreas produces more and more which, over time, can lead to insulin resistance and eventually to Type 2 diabetes and other diseases.

Some preliminary evidence suggests sedentary behavior may have an impact on our mental health as well. Studies of children have found less sedentary kids have better academic scores and higher self-esteem, especially impactful in these days of computer and video games, texting, and surfing the Net. And there seems to be a link between inactivity and depression in adults.

HEART DISEASE

When you sit, blood flows more slowly and muscles burn less fat, which makes it easier for fatty acids to clog your heart. Prolonged sitting has been linked to high blood pressure and elevated cholesterol, and people with the most sedentary time are more than twice as likely to have cardiovascular disease as those with the least. Studies on humans and animals also suggest being idle can lead to harmful changes in your metabolism prompting higher blood fat levels and lower levels of "good" (HDL) cholesterol.

Research published in the *Journal of the American College of Cardiology*, for instance, showed that women who sit for ten or more hours a day may have a significantly greater risk of developing heart disease than those who sit for five hours or less.

CANCER

Findings presented at the 2015 Inaugural Active Working Summit also found that sitting increases lung cancer by 54 percent, uterine cancer by 66 percent, and colon cancer by 30 percent. Another reason for this increased cancer risk is thought to be linked to weight gain and associated biochemical changes, such as alterations in hormones, metabolic disorder, leptin (called the obesity or weight control hormone) dysfunction, and inflammation—all of which promote cancer.

MUSHY ABS

When you stand, move, or even sit up straight, abdominal muscles keep you upright. But when you slump in a chair, they go unused. Cramped back muscles and wimpy abs form a posture-wrecking alliance that can exaggerate the spine's natural arch, a condition called hyperlordosis or swayback.

POOR CIRCULATION IN LEGS AND HIPS

Slow blood circulation also causes fluid to pool in the legs. Problems range from swollen ankles (edema) and varicose veins to dangerous blood clots called deep vein thrombosis (DVT).

Flexible hips help keep you balanced, but chronic sitters so rarely extend the hip flexor muscles that they become short and tight, limiting range of motion and stride length. In the elderly, decreased hip mobility is a leading cause of falls.

LIMP GLUTES

Sitting requires your gluteus maximus ("glutes" are what you sit on) to do absolutely nothing, and they get used to it. Soft glutes hurt stability, making it difficult to get up and to maintain a powerful stride. Sitting down after you've eaten causes your abdominal contents to compress, slowing down digestion. Sluggish digestion, in turn, can lead to cramping, bloating, heartburn, and constipation.

WEAK BONES

Weight-bearing activities such as walking and running stimulate hip and lower-body bones to grow thicker, denser, and stronger. Scientists partially attribute the recent surge in cases of osteoporosis to lack of activity.

FOGGY BRAIN

Moving muscles pumps fresh blood and oxygen through the brain and triggers the release of all sorts of brain and mood-enhancing chemicals. Everything slows, including brain function, when your body is sedentary for too long. New studies indicate that exercise, the more intense the better, actually can keep the mind acute for an extra ten years.

STRAINED NECK

Most sitting occurs at a desk at work, where it is also common to crane the neck forward toward a keyboard or to tilt the head to cradle a phone. Sitting puts more pressure on your spine than standing, and the toll on the health of the back is even worse if you're hunched in front of a computer, all of which can strain your cervical vertebrae and cause permanent imbalances leading to neck stress as well as sore shoulders and back.

It's estimated that 40 percent of people with back pain have spent long hours at their computer each day. The disks in your back are meant to expand and contract as you move, which allows them to absorb blood and nutrients. When you sit, the disks are compressed and can lose flexibility over time. Sitting excessively can also increase your risk of herniated disks.

GET UP AND GET MOVING!

Alter posture by taking light walks to alleviate possible musculoskeletal pain and fatigue and get the body used to more movement. Setting a starting goal of 500 steps a day (which is just about a quarter mile) can go a long way toward getting more movement and less sitting into your life.

At least try to get up once per hour to take a quick walk around the hallways. If you drive to work, park your car at the farthest end of the parking lot and walk all the way to the office. You'll feel less stiff and have more energy throughout the day.

Take breaks several times throughout the day to stretch different parts of your body. If you have a dog at work, take a walk or play.

Regularly break up seat-based work by standing or use adjustable sit-stand desks and workstations. Standing desks have become popular recently, but they aren't new—Leonardo da Vinci, Benjamin Franklin, Winston Churchill, and Ernest Hemingway all used them. Sitting on an exercise ball during the day can aid in working you abs, butt, and legs.

Extend your hands back and rotate your wrists to alleviate stiffness from repetitive tasks (a cause of carpel tunnel syndrome), and also do standing stretches to keep your arms and legs limber.

MORE TO LIFE THAN WORK

You've heard of the expression "get a life." Do it. Don't make a job or profession your whole life. Try to have other outlets outside of the office and spend quality time with family and friends. Keeping photos or mementoes of loved ones at work can be a great morale-booster and a reminder of what probably matters most.

Make an effort to get as much as 8 to 10 hours of sleep everyday—no matter how busy or hectic the schedule. As many as 60 million Americans don't get sufficient sleep. Go to bed and wake up at the same time. Don't drink a lot of caffeine or alcohol during the day, especially before bed. Avoid eating, watching TV, or reading in bed as these things may only make sleep more difficult.

Work and fun are not are mutually exclusive. There are many ways to incorporate fun into the workplace and the benefits can skyrocket employee engagement and output levels. Cut down work tension and pressure by encouraging office sports pools, games, and events, and take the team out occasionally for a drink, ice cream, or pizza.

EAT WELL

Everyone is tired of hearing that breakfast is the most important meal of the day, but it really can help maintain your energy—and your mood. If you don't have time to cook a sit-down breakfast before you leave the house, grab a healthy granola bar or a piece of fruit on the way to work, and keep a few packets of instant oatmeal in your desk for emergencies. Never skip meals as it can raise stress levels, increase grouchiness, and overly strain your system.

When the afternoon slump hits, avoid the chocolate bar, salty treats, or a sugary energy drink as they can cause unwanted side effects such as irritability, anxiety, and "energy crashing." Choose fruits and vegetables that will give you a more natural energy boost.

CLARIFY GOALS AND RESPONSIBILITIES

Every employee has a set of goals and responsibilities for which they are accountable. Chances are, there is room for improvement. One of the biggest contributing factors for falling levels of engagement is the confusion over one's role and how one fits in with and contributes to a company.

Describing objectives and responsibilities is essential to improving employee engagement. A staff member can't be engaged in something they don't understand or for which they don't have the tools. Whenever a project is delegated, it's essential to go into the logistics in as much detail as possible, answering any questions and explaining things completely and clearly.

Learn to prioritize. If you're feeling overwhelmed by a huge to-do list, breaking your workload into individual tasks makes it seem less intimidating. It's great to be a firecracker at work, but be careful not to blow out. Think carefully about taking on additional projects, set realistic goals, and don't try to be a super person.

JOB FLEXIBILITY

The best managers are those who hold their employees accountable for their performance yet give them the flexibility to do their best. The worst are helicopter-control-freak bosses that hover, micromanage, and look over the shoulders of their "underlings."

It's important for the worker to realize just how much flexibility can be utilized in the performance of their job. There are lots of methods to demonstrate that management is willing to grant the flexibility workers need. Allowing employees to work from home is an example, another is permitting flextime or shared projects so long as they fulfill the assignments they're supposed to accomplish and can handle the freedom of flexibility.

ADVANCING CAREERS OR EDUCATION

Empower your employees to take charge of their development. An ambitious employee is one who is driven and focused on improving his or her own career as much as possible. Clever employers inspire employees to improve talents that advance and expand job skills. When managers invest in their employees and help them get better at what they do, they'll be much more likely to be engaged and improve performance.

It's management's duty to offer opportunities for growth and new challenges. This can be achieved in a number of ways including assigning more responsibilities, offering opportunities for advancement such as helping with tuition, encouraging and allowing employees to attend workshops, creating training and mentoring programs, arranging internships, and establishing in-house communication about new opportunities within the organization, especially cultivating and managing opportunities for high-potential employees.

ENCOURAGE NETWORKING

Humans are social animals who share ideas and work together even when it's not required. Ours is an incredibly connected world and allowing and even encouraging staff members to use social media at work inspires them to take ownership of the company's content and to share it across their links.

Network with others (both inside and outside of your organization) and make sure that they have a good understanding of objectives and goals that will aid in a social media strategy. Having goals, perhaps even some safeguards, in place will help ensure that your employees stay on task while they're using social media during working hours.

Send a reminder when there's something new on social media that you'd like them to promote. Plus, this offers workers the chance to exchange information with other departments and businesses.

HIRE INSIDE OR OUTSIDE?

For employees concerned with upward movement within an organization, it's discouraging working for a company that tends to hire from the outside. Almost half of new hires either quit or are fired within their first eighteen months on the job, with 89 percent leaving because they're not a good fit in the company's culture.

Some experts estimate that it costs as much as a half-year's salary to recruit just one employee, given the cost of advertising and training, from the time it takes to screen, interview, and get the new hire on point. A new employee may take many months to get comfortable working in a team setting. Choosing new hires from the outside can upset the corporate culture, leading to breakdowns in team cohesion and efficiency. For the most part, internal promotions are a better way to develop teams within an organization.

When hiring an external candidate, all a company has to judge them by is a résumé, the interview, and perhaps some examples of their work. This puts those hiring at a disadvantage with an unknown quantity applying for a position.

Announcing to a group of employees that the position has been filled externally can hurt morale just as much as the news of a hiring from within can help it. Knowing that all of their hard work will lead nowhere except a paycheck impacts staff effectiveness and loyalty, costing the company money in the long-term as well as diminishing employee engagement in the present.

However, when you promote a current employee, he or she is already familiar with your company's goals and a majority of the tasks associated with its success, and has already demonstrated loyalty to the company's cause and mission.

On the other hand, promoting from within can lead to competitiveness among staff members vying for the same position. When one person is promoted, negative feelings may result in the people left behind. This can lead to dissatisfaction, poor work performance, and resentment from employees who feel slighted or overlooked; it can also cause disruption and confusion for employees who were peers one day and find themselves in a superior-subordinate relationship the next.

Then again, staff already entrenched in the company often lack a fresh perspective that only an outsider can bring. Sometimes employers have fewer options to fill a spot when they look only within the organization. So, think carefully when hiring or promoting.

THE CHANGING CODE OF DRESS

These days people wear flip flops, shorts, t-shirts and tank tops anywhere and everywhere. An employer does not expect employees to be suit-and-tie guys or glam gals but a dress code, or lack of one, can have a big impact on attitude and being comfortable at work. Tube tops and Daisy Dukes might not be the answer, nor are three-piece suits, but reasonable options can have a dramatic impact on the look, feel, and impact of the work environment.

As long as clothes are respectable and work-appropriate, employees should have some leeway and freedom to express individual style and personality. Some feel a more formal dress code fosters professionalism and hard work, while others refuse to believe wearing jeans and sandals somehow detracts from job performance. But people on both sides agree that an uncertain dress code can breed confusion, which can lead to uncomfortable problems de couture.

One recent poll found that most employees are satisfied with their company's dress code policy. However, older workers believe that the dress codes in their workplaces are too lenient, maintaining that low-cut tops, ripped jeans, bare feet, exposed tattoos, and body piercings are not appropriate for an office setting. Also some people do make assumptions about others by the way they are dressed in the workplace. So keep Friday casual, but check out, chat about, and respect employees' feelings about proper dress in the workplace.

USE EDUCATIONAL TOOLS

- The government offers ideas and courses on "Wellness at Work" from the Centers for Disease Control and Prevention (http://www.cdc.gov/features/workingwellness).
- The National Institute for Occupational Safety and Health (NIOSH) offers a business case development fact sheet "Workplace Health and Wellbeing Programs."
- SHRM Foundation's Effective Practice Guidelines Series Promoting Employee Well-Being (https://www.shrm.org/about/foundation/products/Pages/HealthWellBeingEPG.aspx).
- The National Wellness Institute (http://www.nationalwellness.org/).

SMALL BUSINESS HEALTH CARE

The Affordable Care Act (ACA) enacted comprehensive health insurance reforms designed to ensure Americans have access to quality, affordable health care. Find out which Affordable Care Act provisions may impact employers with fewer than twenty-five employees and up to fifty employees (https://www.healthcare.gov/small-businesses/). If you are self-employed, visit https://healthcare.gov to learn more.

SMALL EMPLOYER LOOKING FOR HEALTH BENEFITS?

The Small Business Health Options Program (SHOP) Marketplace helps small businesses provide health coverage to their employees. The SHOP Marketplace is open to employers with fifty or fewer full-time equivalent employees (FTEs), including nonprofit organizations. Learn more about SHOP in your state by visiting healthcare.gov. For questions about SHOP, call 1-800-706-7893 or 1-800-706-7893 FREE (TTY: 711) Monday through Friday, 9 a.m. to 7 p.m. ET.

The Internal Revenue Service is responsible for tax provisions of the Affordable Care Act that will be implemented during the next several years. You can find a list of provisions now in effect, with periodic updates, from the IRS.

THE BOTTOM LINE

Corporate health programs require planning but they ultimately save money while protecting a company's most valuable assets—its employees' health and energy.

5

Scheduling the Staff

Job-Sharing and Alternative Agendas

FACTS AND FIGURES

Alternative job scheduling is a very popular perk among employees, but it can be a pain for managers. Like it or not, it is being used by more and more companies as an incentive for both new hires and seasoned vets. The days where every team member works nine-to-five are quickly fading.

There's a growing body of evidence that the 40-hour plus workweek makes workers less productive, exhausts employees, and can actually be bad for businesses. A recent Harris Poll survey found employees only spend 45 percent of their day on primary job duties. According to the Centers for Disease Control, overtime among industrial workers raises the rate of mistakes and safety mishaps by 61 percent. Longer hours result in lower scores on cognitive performance tests. In other words, longer hours are literally working employees stupid. In fact, employee output falls sharply after a 50-hour workweek, and falls off a cliff after 55 hours—so much so that someone who puts in 70 hours produces nothing more during those extra 15 hours except exhaustion, according to a study published by Stanford University.

This is why many people at work get tired or bored and end up being truly productive for only half the day or less. This is also one of the reasons that some workers sit at their desks chatting online, playing games, gossiping on social media, watching the latest episode of a TV program, or planning for the weekend.

Yet we are now collectively working longer hours. In 1968, households with two working parents put in 53 hours of work a week. By 2000, they worked 64 hours a week. The renowned economist John Maynard Keynes predicted, "By 2030, we'll be working as little as 15 hours a week," and in

1956 Vice President Nixon agreed, "The 4-day workweek is inevitable in 5 to ten years." Alas, employees now work an average of four more weeks a year than they did in 1979.

Longer hours have also been connected to absenteeism and employee turnover. The CDC has an entire Web site devoted to the effects of working long hours. According to US National Health Interview Survey data from 2010, almost 20 percent of adults working 48 to 60 hours or more per week have been associated with health and safety risks.

Some 30 million people work from a home office at least once a week. And that number is expected to increase by 63 percent in the next five years, and 79 percent would like to try at-home work according to a study by the Telework Research Network. People are placing more focus on working remotely than ever before, and CEOs and managers are doing everything they can to accommodate the shift without losing too much control. It can be tricky to find the right amount of balance for out-of-office workers, but it's worth looking into as employees generally find it desirable.

Consequently, more and more employers are offering alternative work schedules, also known as variable work hours, which comprise these main designs: flextime, compressed work weeks, staggered shifts, job-sharing, and telecommuting.

A person who telecommutes is known as a telecommuter, teleworker, or a home-sourced or work-at-home employee. Telecommuting, telework, or remote work refers to flexible work arrangements in which employees—on a regular, predetermined basis—spend all or a portion of the workweek away from the traditional workplace.

In the United States, 37 percent of the workforce telecommutes and typical telecommuters are older employees, working mothers, students, adults caring for elderly dependents, or dual-career individuals, many of whom are seeking to augment their income while starting up their own businesses.

Telecommuters can save between $2,000 and $7,000 a year in expenses, and the greenhouse gas reduction is the equivalent of taking the entire New York State workforce permanently off the road.

If those with compatible jobs and a desire to work from home did so just half the time, the national savings would total over $700 billion a year and a typical business would save $11,000 per person per year, according to Global Workplace Analytics. Telecommuters, on average, save about 29 to 32 miles (each way) per telecommute. Working at home at least once a week will save 20 percent in gasoline costs and wear and tear on your vehicle

When employers staff a position on a full-time basis, job-sharing is an option. This form of part-time employment—in which one position is filled with two (or more) part-time employees who share the duties, responsibilities,

salary, and benefits for one full-time position—the company benefits from
having the special skills and abilities of unique individuals. Each job-sharer
can work up to 32 hours per week and usually the position is prorated between
the two individuals and their total hours worked per week do not exceed 37.5
to 40 hours. Job-sharers are considered part-time employees if they are scheduled to work at least half-time (18.75 hours per week for a 37.5 hour job).
Job-sharers who work less than half-time, by either working fewer than 18.75
hours per week or less than half a year, are considered intermittent employees.

Flextime is a system wherein employees choose their starting and quitting
times from a range of available hours and schedules that allow workers to
alter workday start and finish times. Flextime typically involves a core period of the day during which employees are required to be at work (between
11 a.m. and 3 p.m.), and a period within which all required hours must be
worked between 5:30 a.m. and 7:30 p.m.

A compressed workweek is when a standard week is compressed into
fewer than five days. The most common variation of this is the 4 to 10 in
which employees work four ten-hour days instead of five eight-hour days.
Employees often appreciate this arrangement as it provides an extra day
at home, thus improving work-life balance. Staggered shifts also enable
employees to establish arrival and departure times other than the standard
9:00 to 5:00.

Expanded leave gives employees greater flexibility in terms of requesting
extended periods of time away from work for child or parental care or for
other personal reasons without losing their rights as employees.

In phased or partial retirement, the employee and employer agree to a
schedule wherein the employee's full-time work commitments are gradually
reduced over a period of months or years. This allows older employees to
continue working on a part-time basis with no established end date.

JOB-SHARING AND RESPONSIBILITY

Prepare a detailed job description that identifies each job responsibility, on a
project, daily, or a quarterly basis, and who will be responsible for specific
duties. Establish a schedule for the job-share that accommodates the needs of
the position and is workable for each of the job-sharing partners.

Consider times and settings that encourage communication. Some job-
sharers find it helpful to share an office or even a desk. Set up communication
procedures that specify how the job-sharing partners will convey necessary
information as well as how other people can communicate information to
each other regardless of who is in at any given time.

Be as specific as possible, and identify how frequently each task occurs. Once initial designations have been made as to who's doing what, continually clarify and communicate with each other about what each of you is working on, especially when things become complex. Posting current responsibilities on a wall chart, where they can be easily seen and modified as necessary, works well for most job-sharers.

In designating who will cover each task, think about whether each partner should have separate and distinct responsibilities, or whether both partners will share responsibilities equally. Some are more comfortable having total responsibility for certain tasks. Others find that sharing everything works well for them and prefer that no one task be solely managed by one partner. Remember, though, each job-sharer has accountability for the entire position whenever or whoever is in the office.

Every individual has particular strengths and preferences for certain job tasks. Try to work with your job-share partner to maximize these preferences for both of you, both good and bad. It's also important to address who will cover important meetings. Since some meeting times cannot be flexible, agree beforehand how schedule conflicts will be handled.

SHARE SCHEDULING

When creating a job-share schedule, consider how much overlap time is necessary. Job-sharers find that the more time they can work together, the better their organization and communication. Try four hours at first, and then taper off once workers are in sync. Be flexible about the ability to overlap hours.

One of the advantages of job-share over full-time employees is that they can make arrangements in advance to cover for each other for vacations, illness, or emergencies, which is a strong selling point to workers. Always communicate scheduling with coworkers so they know who will be available for help, consultation, passing messages, and where and when a job-sharer can be contacted.

GETTING GEARS TO MESH

It can be stressful at first sharing an office or a desk. A good working relationship and connecting with a job-share partner takes more than good communication, it also means getting to know each other's styles. Be up front about any issues and get straight with each other as soon as possible. Messages, filing, and neatness habits are just a few of the things you have to learn to coordinate together in addition to sharing the responsibilities of the job. Remember, the primary objective is to accomplish the work in the most efficient way possible.

ACCOUNTABILITY AND COMMAND

Sharing accountability can be tough so talk about how this will happen, and about how job performance objectives will be shared and met, singly or jointly. These relationships need to be established at the beginning of the job-share.

If job-share partners jointly supervise other staff, it's important to be sensitive to the employees who report to you. It's tough to have two bosses as well as for bosses to share employees. Decide about taking on the role of "primary supervisor" so that the subordinates are clear about any change of command. Keep a schedule on who is doing what, when and how it's supposed to get done.

Don't play managers and workers against the other—it's a lose-lose situation. Job-sharing is tough enough without adding potential blockages. Immediately create ground rules that will prevent that from happening.

Internships, for the most part, can be considered alternative work(ers) as most interns don't get paid or have the opportunity for advancement, and much of the time the work is menial. According to the Department of Labor, internships are only legal if they are for the benefit of the interns, not the company. Both the intern and the boss understand it's an unpaid position meant to provide instruction similar to what a person could learn in an educational environment. An internship does not displace any regular employees and offers no promise of payment or permanent job placement in the future. A new study suggests that unpaid interns have almost the same hiring rates as people who have never interned. Unfortunately, the fact is that some companies misrepresent or exploit the relationship so when working as an intern, it is wise to ensure that your workplace is not taking unfair or illegal advantage of you in the course of your assignment. When interning, do not be satisfied with a simple "go-fer" position and ask about tasks that will be educational and will aid you in securing a professional position.

As far as freelancing is concerned, companies regularly work freelancers and independent contractors illegally. Neither can receive unemployment benefits, even though an employer often has requirements—such as shift times, meetings, supervision, office attendance—which technically categorizes these workers as employees.

Always ask to work under the direct supervision of a manager, to receive specific training for any position, to be notified of the times that you have to be available, and to be treated as staff. This will give you leverage if you are treated unfairly, fired, or laid off. By avoiding employee taxes, unemployment insurance, workers' compensation, and healthcare costs, a company can save more than 30 percent of employee expenses, let alone the fact that the United States is the only industrialized nation that does not guarantee health benefits, a paid vacation, paid sick days, and maternity or paternity leave.

CHOOSE CAREFULLY

It's critical for success to match personalities as well as job skills. According to a study conducted at the Institute of Governmental Research at the University of Washington, employers should look for employees who can work independently, are able to set priorities, have good communication skills, and are willing to be flexible about being consulted during their off time.

JOB-SHARE CONS

Despite all the benefits, and while most businesses would love to be flexible with their employees, there are drawbacks to alternative schedules. Without face-to-face interaction and set hours, organizations run the risk of communication breakdown. Employees at companies with flexible work hours may have trouble getting in touch with colleagues, making it harder to coordinate projects, meetings, or phone calls. Leaders may also find it difficult to supervise employees working at different times and locations and this can become a management headache.

Because flexible scheduling is based on trust, less-motivated employees could use a lax, unsupervised environment as an opportunity to work on a tan instead of taking care of business, and it's much harder to hold people accountable when they aren't in the office every day.

JOB-SHARE PROS

Alternative work policies can save companies time and money by reducing demand for office space, facilities, and resources, and employees that work from home use their own office machines increasing their lifespan, decreasing employee overtime. It can make time for emergencies, family, personal time, illnesses, and further education.

The use of employment alternatives can attract or retain highly qualified employees or those with special skills who may not want to work a full-time schedule, it offers more control and freedom, allowing more time for workload surges, and it reduces employment expenditures when employees voluntarily reduce their work schedules.

Other benefits to both employers and employees include lower turnover rates and sick leave, increased job satisfaction among job-sharers, reduced employee stress, phased retirement options, reduced training time and expense (job-sharers train each other), decreased dependence on temporary help (job-sharers ensure assistance), decreased turnover and absenteeism, increased personal time, and the ability to balance work and personal needs.

One company profiled by the Bureau of National Affairs noted that the annual absenteeism rate in their job-sharing program dropped from 4.5 of scheduled work time to less than 0.4 percent.

Employers are also using these job-sharing and scheduling alternatives as powerful tools for recruiting new employees.

6

Air Supply
Heating Up and Chilling Out Air Quality

Fewer things will get you as hot under the collar as being uncomfortable at work. No one should be forced to go "thermal," suffering as your shirt or blouse blooms patches of sweat, or be forced to break out skiing gear to remain comfy.

The Occupational Safety and Health Administration (OSHA) provides guidelines about the most common workplace grumble—its temperature. Lighting and office equipment produce a lot of heat—they account for up to 15 percent of your air conditioning energy costs and greenhouse gas emissions. Many buildings are impacted by what goes on inside the building just as much as, and sometimes more than, the exterior weather conditions.

Today's air conditioners use 30 to 50 percent less energy to produce the same amount of cooling as air conditioners made in the mid-1970s. However, according to the Carbon Trust, air conditioning an office for one extra hour a day uses enough energy in a month to power a TV for more than a year. Air conditioned buildings use roughly twice as much energy as naturally ventilated ones. Even running a fan overnight uses enough energy to power your iPhone for twenty-five years and your laptop for a year.

Finding solutions to large problems can start small. For example, many office buildings can benefit from quick low-cost or no-cost energy-saving solutions such as turning things off, turning things up or down, and following a cleaning and maintenance schedule that keeps equipment running efficiently. Some of these easy actions can yield considerable savings, some at little no or cost at all.

And with automated systems, comfort value is enhanced. Walking into a freezing cold or steamy hot workplace can demoralize the staff and reduces employee productivity from the very beginning of the day. With automation, set the air conditioning or heater to certain temperatures before employees arrive. When they are met with comfortable conditions, they can concentrate on work instead of warming or cooling themselves. Also more money is saved by keeping the thermostat digitally controlled and timed throughout the workday.

FACTS AND FIGURES

Of the roughly 80 billion square feet of total commercial floor space in the United States, about 82 percent is heated and 61 percent is cooled. And energy demand for commercial buildings grows about 50 percent faster than for residential buildings.

Heating, ventilation, and air conditioning (HVAC) climate-controlled systems account for approximately 40 percent of the electricity used in commercial buildings. Improved heating and cooling performance along with substantial energy savings can be achieved by implementing energy-efficiency measures. EnergyStar appliances are 20 to 30 percent more efficient than older models.

The energy costs for an average office building (approximately 20,000 ft.), according to EnergyStar, exceeds $70,000 per year. Reducing that cost can mean significant savings to the bottom line. The United States Office of the Inspector General claimed that businesses could save up to 40 percent for heating and cooling costs for improving HVAC methods in some of the buildings it inspected.

Air leakage accounts for 5 to 40 percent of the costs to heat and cool buildings. Chilled air leaking out of an AC unit can cost $200 (per unit) per year in wasted energy. Leaks can also cause infiltration of airborne pollutants, including pollen, mold, and outside pollution.

AIR QUALITY

Indoor air quality (IAQ) defines the health of the inside of a building and how it can affect workers. Air quality tests should monitor the levels of things like humidity, mold, water damage, type and quality of insulation, asbestos or other toxic substances.

The Occupational Safety and Health Administration (OSHA) has specific IAQ standards and although there are no specific tests for IAQ complaints, the General Duty Clause of OSHA requires that workers have a safe workplace that does not have any known hazards that could cause illness, injury, or death and should include proper ventilation, humidity and temperature

Table 6.1. Minimum Outdoor Air Volume per Person

Building/Room	Cubic Feet per Minute	Liters per Second
Classroom	10	5
Office	5	2.5
Laboratory	10	7
Multi-Use	7.5	3.8

Source: US Environmental Protection Agency.

control, management of pollutants from inside and outside buildings, and an adequate supply of clean, fresh air.

Watch for respiratory or skin symptoms that occur at work, other people in the building with the same complaints, and symptoms that get worse or do not improve. Frequently monitor readings of CO_2 and radon sensors in the building.

Watch for:

- Headaches, fatigue, trouble concentrating, and irritation of the eyes, nose, throat, and lungs.
- Heavy sweating.
- Weakness and/or dizziness.
- Cold, pale, and clammy skin.
- Fast, weak pulse.
- Cramps.
- Nausea or vomiting.
- Fainting.

If anyone presents with these symptoms:

- Move the person to a more comfortable location.
- Lie down and loosen clothing.
- Apply cool, wet cloths to as much of the body as possible.
- Sip cool, not frigid, water.
- If symptoms persist, get medical aid.

According to regulations, if employees suspect a problem, they have a right to contact an OSHA office (check their website (https://www.osha.gov/). The toll-free number is 1-800-321-6742 (OSHA). Anonymity and privacy will be honored.

Eliminating impure sources is done, for example, by finding and eliminating mold and mildew and by stopping any condensation in building materials and HVAC systems, by installing floor grates to avoid dirt and other detritus being tracked into the building, and by choosing interior finish materials that do not "off gas" volatile organic compounds (VOCs).

Some advanced (smart) building information management systems adjust the rates of incoming outdoor air to keep indoor air fresh enough and clean enough without using too much energy. These control systems use CO_2 sensors to test the freshness of air and advanced ones even use pollutant sensors to measure air cleanliness.

Each room should be ducted, avoiding return systems that draw air from one room to another, and don't locate air supply or return registers close to one another.

CHILLING OUT

Replacing an older unit with a newer or more efficient air conditioner can save 20 percent of cooling energy costs. Depending on the season, set thermostats at between 65 and 78 during business hours. Trees can attractively shade the facility and help clean the air. Interior curtains or drapes can be useful in both summer and winter by helping to hold heat and chilled air.

Save 10 to 20 percent of cooling costs by setting the temperature to above 80 degrees Fahrenheit (26.66 degrees Celsius) before you leave the office during summer months. Opening windows and using fans instead of power-hungry air conditioning can save 2 to 3 percent of cooling costs while still feeling comfortable and saving energy.

During cooling season, block radiant heat gain from the sun with reflective window film on the east and west sides of the building and use solar screens, awnings, blinds, and vegetation to keep it cooler. If air conditioning is necessary, be sure windows and doors are shut and close the curtains.

A damper AC vent, called an economizer, draws in cool outside air when available, reducing the energy needed to run an air conditioner. It is made of a valve or plate that stops or regulates the flow of air inside a duct, chimney, or variable air volume box (VAV), a type of heating, ventilating, and air conditioning (HVAC) system. Unlike constant air volume (CAV) systems which supply a constant airflow at a variable temperature, VAV systems vary the airflow at a constant temperature. A damper may be used to cut off central air conditioning (heating or cooling) to an unused room, or to regulate it for room-by-room temperature and climate control. Its operation can be manual or automatic. But be careful! An economizer stuck in the fully opened position may add as much as 50 percent to a building's annual energy bill. Remember, fans can cool an area as much as 3 degrees during the summer and are inexpensive to run.

In dry climates, consider evaporative (swamp) coolers that use the evaporation of water to cool spaces, eliminating the need for energy intensive compressors. In humid climates, consult your HVAC professional about supplemental dehumidification. By controlling humidity at your facility, you can increase occupant comfort and allow for further downsizing of air conditioning equipment.

To check AC efficiency, use a thermometer to measure the temperature of the return air going into the AC unit, then check the temperature of the air coming out. If the temperature difference is less than 14°F or more than 22°F coming out, it may be a problem—have a HVAC professional check the system.

When replacing air conditioning units of five tons or greater, purchase units with a high Energy Efficiency Ratio (EER) rating of 11 or more to reduce operating costs. In larger facilities with energy management systems, verify that temperature settings and operating schedules are correct.

HEATING UP

Set the temperature to between 60 and 65 degrees Fahrenheit (15 and 18 degrees Celsius) before you leave the office during winter months.

To use and control direct sunlight (solar heat gain) to warm your rooms, open blinds on south-facing windows during the day to allow sunlight to naturally heat your workspace. At night, close the blinds to reduce heat loss. Remove solar screens, blinds, or awnings on the south and west facing windows to help increase heat gain during the winter months.

Use portable fans when possible as they circulate the warm air, pulling it away from the ceiling. Remember to change ceiling fan rotation. During winter, blades should run in a clockwise motion; rotate them counterclockwise in summer (most fans will have a switch or pull chain to change rotation). Keep vents closed in unoccupied areas to avoid wasting energy on storage areas and closets. Use directional vents to position warm air currents.

AUTOMATION

To avoid thermostat wars, think about using a wireless control network for HVAC via sensors and routers to and from the thermostat so that there is a single building control system, eliminating the cost and hassles of separate control systems. These systems can be programmed for each day, each season, and special needs so that the thermostat is out of the hands of people who constantly raise and lower temperatures. This lowers cost, saves energy, and avoids thermostat skirmishes.

Keeping air in circulation can help offices feel more comfortable and cooler by creating a wind chill effect. Discuss treating sunlight-rich windows with sunlight-blocking film or by adding blinds. Reducing the amount of sunlight infiltrating these offices could reduce the need for air conditioning use to make them comfortable. Read about other available devices that can be integrated into a computer, smart, or mobile app control source in chapter 11, "Smart Buildings."

APPROPRIATELY SIZED EQUIPMENT GUIDELINES

When purchasing room air conditioners or light commercial heating and cooling units, select those units that are EnergyStar-qualified and reduce building needs by using newer, smaller, and more efficient heating and cooling systems.

Consider energy recovery ventilation systems to reclaim waste energy from the exhaust steam (from boilers) and use it to condition the incoming fresh air. For facilities that have heat-generating processes such as cooking, consider heat recovery as a way to capture free waste heat and use it to offset facility heating and cooling costs.

During the warmer seasons, the system precools and dehumidifies air; it also humidifies and preheats air in the cooler seasons. The benefit of using energy recovery is the ability to meet the ventilation and energy standards of the American Society of Heating, Refrigerating, and Air-Conditioning Engineers (ASHRAE) while improving indoor air quality and reducing total HVAC equipment capacity.

This technology has not only demonstrated an effective means of reducing energy costs in controlling heating and cooling loads, but has allowed for the scaling down of equipment. Additionally, this system will allow for the indoor environment to maintain a relative humidity of 40 to 50 percent in all conditions. The only energy penalty is the power needed for the blower to overcome any pressure drops in the system.

The function of an economical air economizer system, whenever the HVAC system is cooling air, is to shut off the compressor and bring in cool outside air to satisfy the needs of the building.

When selecting a new cooling system, a HVAC professional can provide a quote and specifications for a standard efficiency and high efficiency unit comparison, including maintenance life cycle costs.

For areas such as warehouses and garages, consider installing radiant heating. Radiant heating warms objects instead of the air and requires less fuel.

Radiant heat is also useful for warming exterior areas such as patios and waiting areas.

Be careful about the size of equipment. Too large or too small can increase the cost at the time of the installation and the costs of operation. Request that a HVAC professional conduct an Air Conditioning Contractors of America (ACCA) Manual N Commercial Load Calculation to ensure proper sizing.

At a minimum, specify National Electrical Manufacturers Association (NEMA) premium motors on HVAC equipment, and consider specifying variable speed drives (VSDs) on condenser and evaporator fans.

When possible, add additional insulation to reduce air leakage. Install energy-efficient windows such as EnergyStar windows. Remember that energy-efficient lighting systems emit less heat into office space than older equipment. Reduce solar heat gain via cool roofing and window tints and take steps in cold climates to increase solar gain.

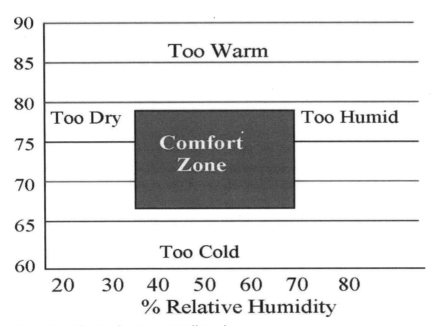

Figure 6.1.   The Comfort Zone. By Jeff Dondero.

THE COMFORT ZONE

Annual surveys by the International Facilities Management Association (IFMA) find that issues of uncomfortable temperatures in the office are primary concerns and complaints. Depending on the season, temperature should be around 68 in the summer to 78 in the winter—plus or minus a couple of degrees depending on personal preferences. Humidity should be between 30 to 60 percent and the maximum airflow should be about 50 feet per minute for cooling and 25 feet per minute for heating.

Relative humidity levels below 20 percent can cause discomfort by drying the mucous membranes and skin. Low relative humidity levels may also cause static electricity build-up and negatively affect the operations of some office equipment such as printers and computers. And relative humidity levels above 70 may lead to the development of condensation on surfaces and within the interior of equipment and building surfaces that can cause rot, mold, and fungi to develop.

PERSONAL DOMAIN

Dress in layers when the office is too cold or warm; that way you can shed or add on to remain comfortable. Imbibe hot or cold beverages to help maintain comfort levels. Engage in less physically demanding activities during peak temperature periods. Use safe and sane work practices to reduce worker exposure to unhealthy temperatures.

Employees have the right to refuse unsafe work; if they reasonably think that stress due to uncomfortable or dangerous temperatures may be hazardous to their health and safety, then a work refusal is possible and permissible.

KEEP IT CLEAN AND HEALTHY

During any construction, keep worker exposure to pollutants to a minimum and ensure that the air distribution system is working properly and is not disrupted. If there are complaints about a specific area, check it out, log changes, and seek reasonable remedies.

Keep air vents clear of papers and office supplies—it can take up to 25 percent more energy to pump air through blocked vents, which can cause a change in temperature and extra energy expended.

Overall, energy consumption can be reduced to less than 10 percent of the average building use through measures such as more and better-grade insulation, passive solar design, low infiltration (drafts), reducing heating and cooling loads, operating restroom fans in connection with the lights, insulating water heaters and supply pipes and ducts, and installing ceiling and wall insulation for comfort—all of which will lower utility bills. Install time clocks or setback-programmable thermostats to turn off systems when the building is unoccupied to maximize efficiency.

MAINTENANCE

Heating and cooling systems require regular annual tune-ups and maintenance that can save from 5 to 15 percent of HVAC costs. Regularly change or clean HVAC filters immediately when wet, and the general rule for air filter changes is that they should be checked every month and changed when they get dirty. A monthly visual inspection allows you to make sure that the filter is still clean enough to do its job.

As temperatures increase, the cooling system runs more frequently. This means that a much larger volume of air will be moving through the system's air filters. At the same time, there will be more particulates hitting the filter, which means it is very likely to get dirty faster.

Also clean all the evaporators, condensers, coils, and heat exchanger surfaces at least once a month, especially during peak cooling or heating seasons. New filters usually only cost a few dollars and dirty filters make the system work too hard which results in lower indoor air quality and more energy use costing far more than a new filter.

Check and repair leaks in piping, air ducts, coils, and fittings, and replace defective equipment, insulation, ducting, and piping. Settings for controls should be checked periodically. Power failures and daylight saving time changes may require settings to be adjusted.

Once a year wet-wipe the following with a one-percent bleach solution to remove dirt and debris:

- Grills.
- Fans.
- Drain pans.
- Ducts and pipe insulation.
- Outdoor air intake areas.
- All belts (and tighten or replace, if necessary).
- Plenums (an air-distribution box attached directly to the supply outlet of the HVAC equipment that heats or cools the air to make rooms comfortable).
- Heating coils, heat recovery coils, heat exchangers, and evaporators.
- Cooling and condenser coils and evaporative coolers.
- Humidifiers.
- Airflow measuring stations.

AIR AID

A Health Hazard Evaluation (HHE) at http://www.cdc.gov/niosh/hhe/ can be requested from the National Institute of Occupations Safety (NIOSH) at no cost.

Contact NIOSH by phone at 1-800-232-4636/Skype: 513-841-4382, Monday through Friday, 9 a.m. to 4:30 p.m. Eastern Standard Time (EST); via email at HHERequestHelp@cdc.gov; or by mail at National Institute for Occupational Safety and Health Hazard Evaluation and Technical Assistance Branch, 1090 Tusculum Ave., Mail Stop R-9, Cincinnati, OH 45226-1988.

7

Illumination

Let There Be (the Right) Light

After air quality and keeping warm or cool, proper lighting is the most expensive item in a building's operating budget. Nineteen percent of energy use in the world is for lighting, and 6 percent of greenhouse emissions worldwide derive from the energy used for lighting. In the United States 22 percent is for lighting.

If every US computer and monitor were shut off every night, we could shut down eight large power stations and decrease CO_2 emissions by 7 million tons annually, and if 10 million office workers turned off unnecessary lights for thirty minutes a day, it would be enough to illuminate 1,312,336 feet of office space, according to the EPA.

Proper lighting is vital for the physical and mental well-being of employees. Besides ensuring that people won't fall down in the dark or stress their eyesight on the job, being stuck in a windowless room under fluorescent lights for hours on end might even be considered torture to some people. Aged fluorescent fixtures can suffer up to 25 percent light degradation over their lifetimes, which can lead to uninspiring lighting and a dull, dusky working environment.

Poor lighting can create a variety of problems, including high human error rates, eye strain, headache, a reduction in mental alertness, and a decrease in productivity, and it can even induce depression and impact the quality of life—all of which contribute to low employee morale and job dissatisfaction.

Ergonomics is an uptown word meaning the art and science of designing workplaces to improve human performance and health. In terms of illumination, it means that good lighting is vital for the employee to see and think clearly, saves money, and creates a healthier, more pleasant workplace—if done correctly.

Figure 7.1.   The Bad Lighting Look. Source: www.giantbomb.com.

Good lighting is really a no-brainer. As much as 50 percent of the initial cost for improvements can be defrayed through practical suggestions. There also are very good products on the market and beneficial luminosity can be accomplished fairly easily and inexpensively.

FACTS AND FIGURES

Lighting is often measured either by the amount of light falling on a surface (illuminance) or the amount of light reflecting off of a surface (luminance). By the year 2030, the US DOE estimates that light emitting diode (LED) technology could save about 190 terawatt-hours of electricity per year. That's enough electricity to power 95 million homes annually. And at today's prices, that's $15 billion in savings, especially attractive as inefficient lighting technology can account for 40 percent of the office electricity bill.

LED lights employ a semiconductor device that converts electricity into light. They are super energy efficient, using approximately 85 percent less energy than halogen or incandescent lighting, which means significant savings on your power bills, plus they have three or four times the lifespan of other types of lighting.

In compact fluorescent light (CFL) bulbs, an electric current is driven through a tube containing argon and a small amount of mercury vapor. This generates invisible ultraviolet light that excites a fluorescent coating (called phosphor) on the inside of the tube, which then emits visible light. CFLs need a little more energy to get started but then use about 75 percent less energy to produce the same amount of light as old-style incandescent bulbs and they last about ten times longer. The electricity used over the lifetime of a single old-style incandescent bulb costs five to ten times the price of a CFL bulb.

The life of one CFL bulb will save the Earth a ton of carbon dioxide and 20 pounds of sulfur dioxide from a coal-burning electric plant, or one and one-quarter barrels of oil from an oil-burning plant, or enough energy to run an average car 1,000 miles.

By following practical suggestions, 50 to 90 percent of lighting energy could be saved without cutting down on illumination or reducing comfort, efficiency, or productivity. In addition, using less electric lighting reduces heat gain, thus saving air conditioning energy and improving comfort and efficiency.

PRACTICAL SUGGESTIONS

Providing the right lighting can save a minimum of 15 percent on the lighting bill, and simply turning off lights and switching to CFLs and lower-wattage bulbs can make a huge difference. By replacing a 100-watt incandescent with an equivalent 25-watt CFL, you can save more than $90 per bulb in electricity costs over the 10,000-hour lifetime of the CFL, and you can reduce energy use by about 75 percent or a savings of about $6 per year for each bulb, according to the US DOE EnergyStar program.

LED lights are "directional" light sources, which means they emit light in a specific direction, unlike incandescent and compact fluorescent bulbs which emit light and heat in all directions. For this reason, LED lighting is able to use light and energy more efficiently in many applications. Making LED-based lighting the primary light source in offices can result not only in a 50 to 78 percent savings in lighting energy, but also in overwhelming user satisfaction. Fifty LED lights can do the work of ninety-six twin-tube fluorescent fixtures with an annual savings of $2,600 or up to $200 for every five bulbs replaced, according to the Environmental Protection Agency (EPA).

Save big without spending a cent by removing one tube from twin fluorescent fittings; this is called "de-lamping." If it's bright enough, de-lamp every second pair of lights. If your facility uses T12 fluorescent lamps, replacing them with T8 lamps and electronic ballasts can reduce lighting consumption by 35 percent. Switch off six pairs of regular fluorescent lights in a conference room an extra twenty hours per week and you'll save 258 pounds of carbon dioxide in a year and $70 off your bill. Adding special reflectors, lenses, and occupancy sensors will double the savings. Switch off fifteen halogen lights in the office one hour earlier at the end of each working day and the annual savings will be around 110 pounds of CO_2 and around $30 in electricity annually.

Install EnergyStar-qualified exit signs. These exit signs can dramatically reduce maintenance by eliminating lamp replacement, and can save up to $10 per sign annually in electricity costs while preventing up to 500 pounds of greenhouse gas emissions from getting into the atmosphere.

Incandescent or halogen desk lamps are handy for lighting personal work spaces, but they are very inefficient. Consider replacing these lamps with CFLs—the newer ones are warm in tone and don't buzz or flicker and they use 25 percent less electricity.

An inexpensive way to increase light is to lower the height of light fixtures to amplify usable light. Color code circuit breakers and mark light switches and that can be turned off when not needed. Utilize natural light including high or clerestory windows, light shelves, and well-placed skylights to help distribute sunlight inside. Paint dark walls and ceilings with lighter colors to maximize the effect of existing lighting.

Install more efficient security and parking lot lighting. High-pressure sodium fixtures are more efficient than metal halide, mercury vapor, fluorescent, or incandescent fixtures.

Use automation and install time clocks or photoelectric cells to control exterior lighting, advertising sign lighting, and some interior lighting. Install motion detectors, dimmers, or sensors to control lighting in frequently unoccupied areas such as restrooms to automatically turn off lighting, and connect fans to operate with lights to reduce energy use and costs. Energy-saving light controls may provide companies with tax credits and many energy utility companies have programs that offer rebates for lighting retrofits.

The EPA Green Lights Program is a partnership program designed to promote efficient lighting systems in commercial and industrial buildings and to prevent pollution, offering organizations the opportunity to use lighting options to cut their power bills in half. If energy-efficient lighting were used wherever profitable, the nation's demand for electricity would be cut by more than 10 percent, leading to 4 to 7 percent reductions in total emissions of carbon dioxide, sulfur dioxide, and nitrogen oxides. In terms of carbon dioxide, EPA finds that this reduction in emissions would be the equivalent of removing 44 million cars from the road! Participants may earn an average 58 percent return on their investment; owners and designers of commercial buildings may qualify for federal tax deductions of 25 to 50 percent to cover the costs of improved lighting equipment.

Engaging in a lighting retrofit can result in reduced energy use and costs by also replacing older fixtures. A lighting retrofit will typically run about $1 to $2 per square foot of office space, including fixtures and installation, and will have a one- to two-year payback period.

Upon becoming a voluntary participant in the Green Lights non-regulatory program, an organization signs a "Memorandum of Understanding" with the EPA which then assists the participating organization throughout the retrofit process, providing lists of manufacturers, lighting management companies, and utilities, called allies, that produce lighting products and provide efficient lighting services. The EPA also provides participants with a computer software program called the Decision Support System that allows corporations to survey lighting systems in their facilities, assess their options, and select the best energy-saving, efficient, and profitable lighting upgrades.

Table 7.1. Lighting Control Strategies

Common	Savings
Dimming: Enable fixtures to dim, for use in other strategies below	Variable
Occupancy Sensing: Adjust lights based on occupancy detection	Up to 40 percent
Scheduling: Dim and turn off lights according to a pre-set schedule	Up to 40 percent
Advanced	
Daylight Harvesting: Adjust electric light levels to take natural light into account using photosensors	Up to 20 percent
Task Tuning: Reduce maximum light levels based on requirements for each space	Up to 20 percent
Demand Response: Reduce light levels at peak times based on automated signals from electric utilities	Variable
Personal Control: Enable individuals to set light levels to suit personal preferences	Up to 10 percent
Energy Management: Software for ongoing improvement in control settings and strategies	Variable
Combined	Up to 70 percent

Source: www.dol.gov/.

LIGHT UP PEOPLE'S LIVES

The ability to control lighting has positive impacts on workers' moods, satisfaction, and productivity. Good lighting can decrease errors by 30 to 60 percent as well as lower insurance premiums and reduce eye strain, headaches, nausea, and neck pain.

Utilize task lighting, which is any localized light source directed on some activity or task (as simple as an adjustable-arm desk lamp). Avoid glare or indirect lighting for detailed tasks as it either creates harsh light or directs almost all light to the ceiling. People prefer lower brightness for computer tasks and higher brightness for reading, writing, or eye-to-hand work.

Most people prefer natural light or daylight which in most types of work settings is an asset. Humans are hard-wired to like the sunlight and using it saves energy, is comfortable as a light source, costs nothing, and should be part of a lighting program whenever possible.

Most workplaces are overlit, too bright, and affect both the employees' productivity and a building's bottom-line energy costs. Visual comfort means ensuring that people have enough light for their activities with the right quality and the right balance—not too harsh or dim— helping to create a secure and productive environment. Having good, lighted views and sight lines gives people a sense of well-being and control of their environment.

Pairing lower levels of ambient light (light that is natural) with task lighting can provide optimal conditions. When ambient and task lighting are layered, direct lighting is brought closer to the work surface reducing the lumens (light output) needed to adequately illuminate a workspace. Lamps that are high in color temperature provide improved visual acuity compared to lower-color temperature lamps at the same light level.

Light bulbs shine in *color temperatures* from warm tones to cool. Low watt, *high* lumen *bulbs* save power and give more brightness. Color temperatures over 5000 Kelvin are called cool colors (bluish white), while lower color temperatures (2700–3000°K) are called warm colors (yellowish white through red).

HANDS-OFF LIGHTING

Anytime a computer-based system can assume a basic but important task, a business is going to run more efficiently, cheaply, and hopefully, with some sideline benefits. When employees don't have to remember to shut off the lights on their way out, turn lights on at a given time according to the amount of sunlight, or pay attention to the special needs of some workers and the occupancy level of rooms, energy will be saved. Consequently, occupancy sensors have become the lighting control of choice for reducing wasted lighting energy in common-area applications.

Building automation systems (BAS) are interlinked, centralized networks of programs that control workplace environments. Lighting controls are considered an easy fix in the BAS field since the expense can be much lower than other elements of energy-saving equations.

A lot of businesses face high electric bills due to lighting rooms or facilities without occupants. The basis of a building automation system operates via the use of smart sensors and controllers that help manage all aspects of the involved network and equipment. In order to function properly, controllers need input from sensors that take information from the environment and send it to the controllers which automatically adjust energy consumption levels. Businesses are finding that computer-controlled light systems are affordable, will save a substantial amount of money over time, and generate fewer greenhouse gases.

However, even good equipment can be installed incorrectly. Don't install sensors behind a coat rack, door, bookcase, curtains, etc. They must be able to "see" an approaching person's motion to turn on the light before they enter or leave an area. Controls are also sensitive to demand response. If a business can reduce its consumption at peak energy demand times, it will often be reimbursed by power companies for doing so. Controls can be set to use sunlight as well, so trim or remove the trees and bushes near windows to maximize exposure to sunlight and to reduce shadows.

Automated systems can even leave certain lights on to make the business appear busy at night, a security byproduct. Additional lighting can also be attached to the security system. Upon a break-in or unscheduled arrival, authorities can be notified, lights turned on, video cameras started, and thieves can be caught in the act.

If luminosity is well done, it also provides an opportunity to earn Leadership in Energy and Environmental Design (LEED) points. LEED-certified buildings are resource-efficient and allocate points "based on the potential environmental impacts and human benefits of each credit." LEED points can be beneficial in a number of ways such as enhanced property values, insurance breaks, and economic incentives from local and federal agencies.

A well-lit building enhances comfort and safety for both visitors and workers. Visitors can safely walk, read available literature, and get around the facility more easily. Lighting plays a crucial role in ergonomics, the term given to overall workplace health and safety.

MAINTENANCE

Computerized systems also offer automated maintenance with lighting elements and a central controller working in tandem to facilitate automatic alerts. This allows for rapid troubleshooting and alerting personnel to remove or lessen an interruption of service.

When lamps, skylights, windows, ceilings, walls, and work areas are maintained in good condition, lighting can be improved without increasing the number of fixtures or light bulbs. Immediately replace light fixtures that have blown bulbs or are reduced in brightness. Clean lamps and shades at least twice a year. Lamps give off less brightness when they are dirty but still consume the same amount of electricity. In a dusty workplace, light coming from dirty windows is reduced 30 to 40 percent after three months and 45 to 55 percent in six months.

LIGHT UP THE FUTURE

When lighting facilities are connected to the Internet, streetlights will be equipped with cameras and sensors that allow city managers to monitor traffic and identify roads that need to be serviced as well as weather conditions, pollution levels, pollen counts, and traffic conditions.

Lighting fixtures may contain speakers for broadcasting emergency warnings or Wi-Fi capabilities. In parking lots people may be able to launch an app tied to the lighting system to find an open parking space. Within retail stores, location-based technology embedded in LED lighting fixtures can give retailers the ability to push targeted items and bargains to shoppers' smart phones or direct them to a desired product.

LED Internet installations will become increasingly more affordable. Smart solutions will use lighting's infrastructure as a conduit to "Big Data" (a term for data sets that are so large or complex that they may be analyzed by computers to reveal patterns, trends, and associations relating to human behavior) to increase usability of information, thereby reducing time spent and consumer costs, very attractive to shoppers.

8

A Vampire in the Workplace

Standby Electricity

FACTS AND FIGURES

It's hard to believe that up to 20 percent of the total energy costs in the workplace represent zip, zilch, zero. Standby "vampire" or "phantom" power is the energy consumed by electronic appliances when they're "hanging out" or waiting to be used. More than half of the energy used by some electronic devices is standby electricity for instant-on capacity—that is, no wait or warm-up time.

This vampire energy can represent a substantial portion of the overall energy use of an office and is reflected in the monthly energy bill that has its fangs in the business budget. Unfortunately, few people probably even know what it is, what's happening, or how much it costs. The bottom line of running a successful business is to make all the dollars count. That should begin with eliminating one of the biggest wastes—power that is paid for but is not productive.

According to the Lawrence Berkeley National Laboratory (Berkeley Lab), standby power use is responsible for approximately 1 percent of global carbon dioxide emissions and annually equals the output of seventeen one-megawatt power plants, which can power more than 6,000 homes.

Almost any electrical device with an external power supply, remote control, or continuous display (including an LED) that is not completely turned off or that charges batteries will draw power continuously. Sometimes there is no obvious sign of constant power consumption and an electronic meter is needed to be certain.

Items left plugged into the wall, such as a mobile phone charger or laptop adapter, can leak more than 20 watts of power each per day. Some of the most prevalent standby power devices are the large plug-in transformers that are used to step down power to cordless phones and answering machines.

STANDBY POWER IS NECESSARY FOR:

- Maintaining a signal reception for remote control and telephone signal networks.
- To receive immediate control of a device.
- Internal clocks (microwave ovens use more power for the clock than for food preparation).
- Battery charging.
- Continuous display of a device's condition (the light that remains on even if a machine is switched off).

Many computers and peripherals like copiers and printers, when left on, consume 25 percent of their power when not in use. About 30 to 40 percent of office equipment is left on at night and on weekends. A monitor left on overnight can use the same amount of energy a laser printer uses to produce 800 copies. Even when idle, laser printers consume between 30 and 35 percent of their peak power requirements.

Most new electronic equipment has low standby power (1 to 5 watts, compared with older products that needed 10 to 20 watts to remain idle). Alternatively referred to as sleep mode, standby is a mode the computer, monitor, or other device enters when idle for too long. This mode helps conserve power when a computer or computer device is not in use without having to sacrifice the time it would take to turn the computer off and on again. A surprisingly large number of electrical products—TVs to microwave ovens to air conditioners—cannot be switched off completely without being unplugged. To find out your equipment's standby power consumption rate, either check the manufacturer's spec sheet or manual, or measure it with a power meter.

One watt of standby power equates to about $1 year in consumption. This may not seem like much, but when you have multiple appliances in the office drawing an average of 5 watts when they're off, this can add up. If you invested the $200 average spent on standby power annually, it would come to more than $24,000 in thirty years (at 8 percent average growth).

Each device may only consume a few watts, but multiply the number of devices over a 24-hour day in a typical office and the scope of the possible savings is humungous for machines that are not even turned on. Remember that every 1,000 kWh that is saved by turning equipment off represents about $120 per month savings (based on $.12 per kWh).

Table 8.1. The Cost of Standby Electricity

Product	Watts Per Hour When On	Watts Per Hour Standby	Monthly Standby Cost	Light Bulbs Powered by Standby
Digital cable box	25	18	$1.45	33
Computer	130	15	$1.21	28
Modem	14	14	$1.13	26
Monitor	70	11	$0.88	20

Source: USEI.

Smart Strip power strips monitor the consumption of six to eight devices and can sense when computers and other devices are on or off. Upon figuring this out, a Smart Strip shuts off the power, eliminating the idle current drawn for them. A Smart Strip power-saving surge protector can save from $1.60 to $5.70 on electricity costs per month and will pay for itself in six months.

In January, the California Energy Commission (CEC) introduced mandatory standby requirements for various electronic devices—the first such obligatory regulations in the world. The new standards require approximately a 90 percent reduction in standby from 5 watts to 0.5 watts.

Researchers suggest that an informed and aggressive approach can reduce standby use by about 30 percent in the near future and it's estimated that it is technically feasible to reduce standby power by 75 percent within the next decade.

PRACTICAL SUGGESTIONS

Switching to sleep or inactive mode can be a major-league power saver. A typical monitor consumes 30 to 140 watts while on, but less than 3 watts while "sleeping." The recommended setting is for your computer to go to inactive after no more than fifteen to thirty minutes of inactivity.

Choose settings that automatically switch the computer monitor into sleep or "power-down" mode with a preset time limit to save as much as $44 per year per computer and as much as one-third of a computer's overall energy consumption. Remember that just because they are switched off doesn't mean that they are not sucking up juice. The simplest way to reduce standby power consumption is to unplug devices (i.e., cell phone chargers, fans, coffeemakers, desktop printers, radios, etc.) when they are not being used.

Screensavers are designed to save your screen from burn-in, not to save energy, turn them off. Coordinate with vending machine vendors to turn off machine lights.

Figure 8.1.   Power Leaks of Office Machines per Day. Source: EIA.

When buying new appliances, ask or look for low-energy standby products (look for the EnergyStar label products have lower standby rates)—unfortunately, few products list their standby power use.

Don't forget to look into the peak load reduction incentives and rebates available from the federal government (3.6 billion is available in rebates, incentives, and demand response programs annually), state governments, and your local utility company. For these rebates, check out http://energy.gov/node/782456/financing/energyincentiveprograms.html.

If you really want to know how much energy your devices use, it's time to buy a voltmeter (they are inexpensive).

9

Workplace Waterwatch
Watch What Goes Down the Drain

Not many people think about water conservation at work—maybe because people figure that they don't use that much, it's someone else's job, and there are lots of other things to consider.

Well, water heating alone can account for 9 percent of the total office energy load. It costs a company to buy water, heat water, and dispose of water through sewage charges. Then there is landscaping, which can put a dent in your water budget if not thought out carefully. So, think "team effort" when saving water—it can have a bigger benefit across the board than one might think, let alone helping to cut back on water waste in a time of drought in the United States.

According to the World Economic Forum's Global Risks Report 2016, the water crisis—resulting from shortages, floods, or droughts—is the number one resource dilemma facing the planet over the next ten years. The United States is currently engulfed in one of the worst droughts in recent memory. Almost half of the country experienced at least moderate drought and in several southwest states the condition is critical. Scientists at Columbia and Cornell Universities predict that this region and the Great Plains are likely to experience the greatest mega-droughts since pre-Medieval times.

In fact, in California, a water utility company is using drones with infrared sensors to locate water leaks in farmlands. For water to be the top concern is alarming because it is so tightly interlinked with other global concerns such as involuntary migration, regional and economic conflicts, production of just about everything we use, legal conflict, as well as a dramatic increase in the cost of energy.

HOW MUCH WATER DOES IT TAKE?

- Approximately 39,090 gallons of water are needed to make a car, including 518 gallons to make each tire.
- 1,800 gallons of water are required to grow enough cotton to produce one pair of blue jeans.
- A whopping 400 gallons of water is necessary to grow the cotton required for an ordinary cotton shirt.
- A single 32-gallon keg of beer takes 1,500 gallons of water.
- It takes 53 gallons of water to make a regular-sized latte.
- A gallon of paint uses 13 gallons of water.
- 1.85 gallons of water is used to manufacture the plastic of one commercial bottle of water; most of that "chi-chi" water comes out of the tap.
- 62,000 gallons of water is needed to produce one ton of steel.
- 1,360 gallons of water are used to mix one ton of cement.
- 24 gallons of water are used to make one pound of plastic.
- 55 gallons of water are required to manufacture a pound of synthetic rubber.
- Close to 40 percent of all groundwater in the United States is used in agriculture, and four-fifths of that is used in the southwest, the green belt which grows from 40 to 80 percent of the greens of the country (depending on the seasons).

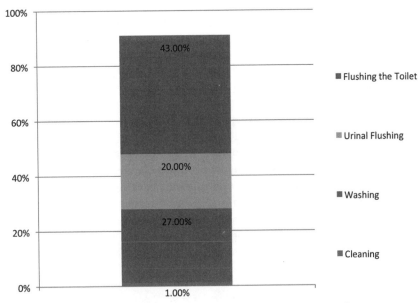

Figure 9.1.   Sanitary Use of Water in the Office. Source: uswaterservcies.com.

TAP THE WATER IN OFFICE BUILDINGS

According to the EPA, using water-efficient appliances at home and work would help the country save more than 3 trillion gallons of water and more than $18 billion per year.

Using water also draws on another precious resource, energy. Letting your faucet run for just five minutes uses as much energy as running a 60-watt light bulb for 14 hours.

When you extrapolate those figures to commercial buildings, which account for almost 40 percent of the country's energy usage, you're talking about a substantial amount of resources that could be saved with simple, efficient practices that might also significantly reduce the operating costs of a business. Taking easy, inexpensive measures can typically reduce your water consumption by up to 50 percent. An auditor can take a look at water use and offer advice on how to be more efficient. Simple measures and fact finding can save 30 to 70 percent of water- and energy-related consumption at the office.

According to the National Resources Defense Council (NRDC), a water audit of your office building includes a review of domestic, sanitary, and landscaping processes and can sometimes be performed for free by your utilities company. Water audits can also be performed by a hired contractor. The NRDC recommends consulting the American Water Works Association's Guide to Suppliers (AWWA), or your office manager can request free water audit software from the AWWA (check out chapter 18 regarding building audits).

Congress enacted the Energy Policy Act (2005) which dictates maximum water consumption requirements for many plumbing devices. The water-using products and appliances covered are:

- Toilets
- Urinals
- Faucets
- Showerheads
- Commercial ice makes
- Pre-rinse spray valves

The Alliance for Water Efficiency (AWE) maintains a list of current and proposed national efficiency standards and voluntary specifications for residential and commercial water fixtures and appliances. This can be a useful reference tool for understanding the types of water-efficient products and appliances that are available for water savings potential.

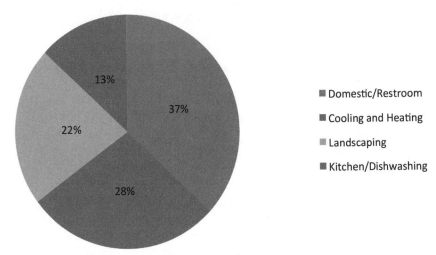

Figure 9.2. Water Use in the Office. Source: bing.com.

Potable (drinkable) water is not required for many industrial uses and can be substituted with impotable or reused water. Sources include but are not limited to air conditioner condensates, cooling tower blowdown, and rainwater.

A recent study by the Ethical Corporation (an organization that specializes in sustainability and environmental responsibility) showed that 99 percent of business managers surveyed ranked water conservation as a "top five" priority over the next decade.

Many employers and employees see drinking water as a health benefit. A recent survey claims that 65 percent of workers want bottled water free at the office, while 40 percent of consumer's desire bottled water from vending machines. In addition, nearly 70 percent of workplaces offer free filtered or jugged water to employees and visitors making free water service second only to free hot coffee at work.

In some places the price of water has doubled in the past five years and will continue to rise. The more efficient your business is now, the more savings you will make conserving water.

The recommended eight glasses of water a day (largely considered erroneous, by the way), at US tap rates, equals about $.49 per year per person; that same amount of bottled water costs about $1,400. Consequently, it makes sense to use tap water and reusable glasses and keep a filtered-water carafe in the break room refrigerator. If you have to use plastics bottles, make sure they are recycled.

Being water smart will help to develop an eco-friendly image, attract environmentally conscious people, and gain access to government support programs. Saving hot water will also save electricity and lower the power bill, so be sure to pay particular attention to any water-use areas that waste hot water.

DODGE THE BOTTLED WATER BILK

Americans spent $11.8 billion dollars on bottled water in 2012, putting the average cost at $1.22 a gallon, and produced more than 2.5 million tons of carbon dioxide in the process of making plastic bottles. Pacific Institute estimates that the equivalent of more than 17 million barrels of oil have been used to produce these plastic bottles and that the total amount of energy for each bottle may be as high as the equivalent of filling a plastic water bottle one-quarter full with oil.

The best-selling bottled water is marketed by PepsiCo (Aquafina), Coca-Cola (Dasani), and Nestle which rake in a combined $110 billion a year worldwide at up to 2,000 times the cost of getting water from a tap, which is where Pepsi, Coke, and Nestle (for some of their products) get their water.

In the United States alone, more than half the population drinks bottled water, which accounts for about 30 percent of liquid refreshment sales, far exceeding the sales of milk and beer; only soft drinks sell more. What's worse is that although more than 92 percent of tap water is regulated and safe, bottled water doesn't have to meet government standards. Municipal water utilities must share their treatment methods and contaminant-testing results with consumers annually; bottled water companies are not required to disclose this information and, in a 2008 study, the Environmental Working Group found 38 pollutants in ten brands of bottled water. In fact Fiji, a very chic H_2O touted as being pure and healthful, was not as pure as tap water from Cleveland.

EMPLOYEE EDUCATION

Management and employees should share a commitment to wise water use by generating and establishing goals through staff meetings, newsletters, and e-mails which encourage employees to drink from the tap, use personalized water bottles, or reuse (non-plastic) cups and glasses.

Post monthly water usage levels. Create suggestions and incentives at your organization to recognize water-saving and conservation ideas. Include water-saving tips in your employee newsletter, featuring how much water can be saved with each suggestion. For instance, communicate that each time a person brushes their teeth without turning off the tap, they waste up to five gallons of water—then multiply that by members in the family, people in the town, county, state, and the nation. Invite a water utility conservation person to speak to your organization.

Retrofitting the workplace with modern water-efficient equipment raises the value of your property. Businesses may also be eligible for governmental rebates under programs like the Metropolitan Water District of Southern California's rebate program. Check your local utility for details.

Have a water audit done for your facility to find out the recommended water use for your operations, then make sure someone monitors your utility bills to gauge your monthly consumption. Show your company's dedication to water conservation by creating a written policy statement and by committing management, staff, time, and other resources to the effort. Create a goal of how much water your company can save and plan a celebration once that goal is met. Provide incentives and reward employees—but not with bottled water—for water-wise comments, use, and follow-through.

WATER-WISE TIPS

- Use WaterSense high-efficiency fixtures, equipment, and appliances.
- Have your system checked for leaks and get them repaired.
- It is important to know how much water is being used for each of your firm's industrial processes and/or domestic needs.
- Collect bills for one year and compare months to see if there are any discrepancies after conserving. Analyze each month's use. Understanding water use will identify savings opportunities, allow appropriate savings targets to be established, and serve as a benchmark from which water savings can be tracked.
- Become a member of Partners for a Clean Environment. PACE recognizes businesses that implement conservation methods.
- Don't use or install ornamental water features unless they recycle water. At least place decorative fountains on a timer and use only during work or daylight hours—evaporation causes lots of water waste.
- Dry sweep or use a water broom when possible, instead of using a hose to clean floors, sidewalks, and other hard surfaces. Water brooms (powerful and reliable for use in pressure washing driveways, boat decks, flat surfaces and more) clean flat surfaces quickly, easily, and evenly and are superior to hose and spray nozzles in both water efficiency and cleaning effectiveness.

- Make sure all hoses are equipped with an automatic shut-off nozzle. Hoses that don't have an automatic shut-off nozzle and are left running can waste 8 to12 gallons per minute.
- Perform tours of the building paying close attention to all water-using equipment indoors and outdoors by listening and looking for unexpected water use. Check monthly water facilities for efficiency: bathroom, toilet, faucets, and shower.
- Look for wet spots suggesting overspray in parking lots, on walkways, and in grassy areas surrounding the facility. When encountering unanticipated soggy spots, contact the water utility to determine if there is a leak.
- Modify or replace equipment or install water-saving devices. Detail recommendations on water-saving technologies, with savings and payback projections.
- Building owners and operators should think about installing a smart water meter data management system with remote capabilities for feedback and to identify problems.

SHOWERS, FAUCETS, WATER HEATERS, AND TOILETS

Toilets and urinals account for about one-third of all water consumed in US buildings. To reduce this cost significantly, install low-flush or timed toilets that use only 1.6 gallons or less per flush—a fraction of the consumption of older models. In many areas, low-volume units are required by law. When the tank is empty, use a gallon bucket to refill the tank to its normal level. If you needed more than 3 gallons of water to fill the tank, consider replacing with a low flush model.

Install toilet dams or displacement devices (a half-gallon plastic bottled filled with gravel or water placed in the tank) to cut down on the amount needed for each flush. The container will save on each flush and the commode will work just as well.

Check your toilets for leaks. Put a few drops of food coloring in the tank. If the coloring begins to appear in the bowl without flushing, you have a leak that should be repaired immediately, because even a small leak can waste thousands of gallons. In fact, a leaky faucet that drips at the rate of one drop per second can waste more than 3,000 gallons per year—or one shower a day for six months. A small, inexpensive, and easy-to-install gasket can solve a big problem.

Check for worn out, corroded, or bent parts. Most replacement parts are inexpensive and easy to find and install. Replace or adjust the toilet flush handle frequently if it sticks in the flush position, which lets water run constantly. Avoid flushing the toilet unnecessarily. Dispose of waste tissue in a trash container rather than in the toilet, which will save water and strain on the sewer systems.

Install low-flow showerheads in shower facilities. A 2.5 gallon per minute (gpm) showerhead will cut energy and water usage from a standard shower head in half, with no reduction in comfort. The new and efficient 1.5 gpm showerheads cut the water and energy use by another 40 percent.

Many small offices have the same 50-gallon hot water tank used in your home. Smaller tanks reduce the "standby" losses from your water heater, and the new tankless water heaters cut standby losses even further. Consider installing an instant water heater so you don't have to let the water run while it heats up.

By adding aerators to your existing office faucets, you can cut water faucet consumption in half. These inexpensive, super-simple little devices are available in most hardware stores, and some utilities offer them for free.

Shut off water to unused areas of your facility to eliminate waste from leaks or unmonitored use. Install water-softening systems only when necessary and save water and salt by running the minimum amount of regenerations necessary to maintain water softness.

When adjusting water temperatures, if the water is too hot or cold, try to mix by turning down rather than increasing water flow to balance the temperature.

WATER- OR AIR-COOLED EFFICIENCY

Water use is often a component of industrial and commercial equipment used for cooling purposes. Often this equipment is also available with technology that uses air for cooling. The pros and cons of each should be determined before switching. A couple of factors to consider are energy efficiency and performance. Cooling towers often represent the largest percentage of water consumption in industrial operations. To improve the efficiency of cooling towers and reduce water use:

- Eliminate once-through cooling.
- Install a conductivity controller on each cooling tower.
- Equip cooling towers with overflow alarms.
- Use high-efficiency drift eliminators.
- Install submeters to monitor makeup and bleed on each cooling tower.

- Properly train and educate cooling tower operators.
- Replace water-cooled equipment with air-cooled equipment when feasible. Newer air-to-air models are just as efficient and do not waste water.

CATERING AND KITCHENS

- A running kitchen tap uses enough to fill a bath in 8 minutes or a minimum of 24 gallons.
- Fit spray head taps, which can reduce consumption by up to 70 percent.
- Fit taps (percussion taps) which cut off automatically. Not only do they save water, they can save staff time and improve safety in a busy kitchen.
- A fully loaded dishwasher uses less water, and is more hygienic.

IN HOTELS

- En suite facilities are a must in today's hotels, but are a major source of water use. However, the largest uses of water in hotels are restrooms.
- Employ "drag strip" or Navy showers—rinse, water off, then rinse again.
- Check the shower flow—a power shower uses up to 10 gallons of water a minute—half of which is fine for most people and you can change it by changing the showerhead.
- Fit water-saving devices to the toilet cisterns, as flushing can be the cause of up to 30 percent of water usage.
- Proper insulation of hot water pipes not only conserves energy, it saves on wasted lukewarm water runoff.
- Educate guests to reuse towels and bed sheets for more than a day.

INDUSTRY

- Steam costs more than ten times ordinary water to produce, so check for inefficient steam pipes as circulation wastes steam as it is vented automatically from taps.
- Consider fitting submeters to check usage in different processes.
- Review the efficiency of your boiler water system.
- Switch off water-cooling systems along with machinery at the end of a shift.
- Only clean windows as required.

LANDSCAPING

- Average landscape water use for the commercial and institutional section can range from 7 to 30 percent for office buildings.
- Have an irrigation professional (or send the person in charge of landscaping to an irrigation conservation workshop) design, install, and maintain the irrigation system, and check regularly that the best management practices are followed.
- For landscaping equipment, install auto stop-leak detection alarm systems on major water-using equipment.
- Regularly inspect the sprinkler heads to make sure they are not damaged or malfunctioning in any way, like spraying water on paved surfaces.
- Drip irrigation can reduce water consumption from 20 to 50 percent and smart systems can further that saving by 15 percent.
- Use native plants or other plants that require little water to thrive in your region. Check out the Plant Conservation Alliance to learn what plants are native to your region. Group plants according to types or by their needs.
- Replace grass with artificial turf or plant grass only in areas where people will actively use it.
- Add mulch to plants and soil to prevent water loss through evaporation.
- Water landscapes in the morning to prevent water loss due to evaporation and wind.
- Think about collecting rainwater in cisterns via drains or "green roofs."
- Never put water or ice down the drain when there may be another use for it such as cleaning or watering a plant or garden. Look into gray water systems that reuse impotable water for irrigation.

WATER CONSERVATION RESOURCES

- WaterSense, United States Environmental Protection Agency.
- Green Business Advisor, National Resources Defense Council.
- A Water Conservation Guide for Commercial, Institutional and Industrial Users, New Mexico Office of the State Engineer (good advice for all for arid states).
- Green Building Resources, United States Green Building Council.
- Energy Efficiency and Renewable Energy Division, United States Department of Energy.
- Local water utilities and nurseries.

FREE WATER AID

If your business is struggling to come up with the initial capital to make a green investment, the US Department of Energy's Energy Efficiency and Renewable Energy division should be your next stop. The EERE has awarded hundreds of millions of dollars in financial assistance to help businesses, schools, and industry upgrade to renewable and energy-efficient technologies. And don't forget to contact your utilities company to learn about available incentives and tax credits for increasing your building's efficiency and to get information about water use of any kind.

10

Building Management

A Safe, Clean, Healthy Place to Work

FACTS AND FIGURES

During our working careers, many of us have labored in drafty, dusty ware-houses, windowless offices cramped with cubicles, or in places where cold can be bitter and heat can be brutal. The one thing all toilers have in common is the need for a comfortable, clean, well lighted, heated, and cooled work-place, with access to some sunlight and clean air.

Think of the workplace or building as a huge machine that has to be con-tinually maintained—a life-support system, communication terminal, data manufacturing center, and much more. They are incredibly expensive tools that must be constantly adjusted to function efficiently.

IMPROVED ENVIRONMENTAL IMPACT OF BUILDINGS IN THE FUTURE

Table 10.1. Improved Environmental Impact of Buildings

Impact Category	Future Level
Total Energy Use	36% less
Total Electricity Consumption	65% less
Total Water Use	12% less
Total CO_2 Emission	30% less
Total Raw Material Use	30% less
Total Waste Output	30% less

Source: EIA.

96

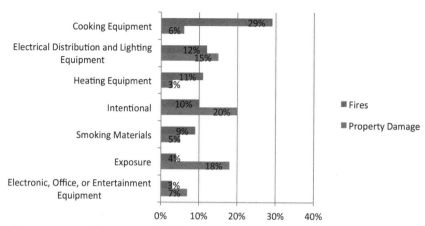

Figure 10.1. Leading Causes of Office Fires. Source: www.cdc.gov.

SUSTAINABILITY IN DESIGN AND MAINTENANCE

We spend nearly 80 to 90 percent of our time indoors and the quality of indoor spaces such as the temperature, humidity, noise, light, space, and air quality affects health and productivity. In fact most interiors, whether at home or in the office, are more polluted than the outdoors.

Buildings not only have impacts on their occupants, but also the environment, society, and economy around them. The buildings in which we work and shop use more than $200 billion worth of electricity and natural gas each year (commercial buildings use $107.9 billion and industrial facilities use $94.4 billion).

The main objectives of sustainability measures are to conserve resources and reduce degradation of the environment, as well as to create a safe and comfortable workplace.

Researchers from Harvard University recently published the results of a study that was designed to simulate indoor environmental quality conditions in green versus conventional buildings and to evaluate the impact on human performance using basic principles of high-performance building design, construction, and operation, and noting their impacts on indoor environmental quality, including chemical exposure, light, noise and thermal comfort.

In kind of a new age "behavioral sink" project, scientists at Harvard cite that overcrowding, air pollution, too much noise, and a lack of access to nature contributes to the emotional and physical stress of urban life. As cities continue to grow, the design of buildings, neighborhoods, and commons should be planned and built to promote human health.

Harvard's Center for Health and the Global Environment encourages decision makers, city residents, and urban workers to understand the importance of natural areas for our environment and health. Suggestions include using sustainable building materials, designs, and techniques, promoting city and building designs that encourage walking, climbing stairs, getting outdoors, and designing urban parks that deliver health benefits and essential ecological services.

Research suggests that even small amounts of daily contact with nature can help us think more clearly, reduce our stress, improve physical and emotional health, and nurture a personal connection to the environment that will compel us to protect it.

As the world's climate changes and if extreme weather patterns become more common, buildings designed to operate under conventional climatic conditions may not perform as expected in future climatic conditions. For example, a building that was made airtight to increase energy efficiency might create problems such as decreased ventilation and more occupants reporting health problems due to the high concentration of many hazardous substances such as volatile organic compounds (VOCs) that can off-gas unhealthy vapors from indoor sources. Such emissions of indoor contaminants can come from building materials, furniture, and cleaning products, and indoor environmental conditions like humidity can alter the levels of microorganisms, insects, particulates, and allergens.

BUILDINGS SHOULD . . .

Creating sustainable buildings starts with good site selection, including consideration of the reuse or rehabilitation of existing buildings. The sustainable and green building deals with "environmentally responsible and resource-efficient" structures, provides comfortable environments, and is engineered to accommodate change to technological tools. This kind of facility will contribute to improved working environments, higher productivity, reduced energy and resource costs, and preventable system failures.

- Above all, a building's systems must be safe.
- Assess infrastructure damage from major natural disasters.
- Take measures to deter, detect, delay, and/or respond to emergencies.
- Be careful what is used for sound barriers—and ensure that they meet fire resistant requirements. Do not use bedding or packing foam, as these can be flammable.
- Ensure that there is safe drinking water.

- Think about directional sound technology that is not language-specific and can guide people to safety exits. It is especially helpful in smoke-filled environments; different tones alert occupants to go up or down stairways.
- Provide visual clues for people with hearing impairments.

To conserve energy, it's important to find ways of reducing the energy load and increasing efficiency while improving energy performance in existing buildings, and keeping tabs on carbon footprints. This ties in with increasing our energy independence and being committed to net-zero energy use goals.

A sustainable building should use water efficiently and, when feasible, reuse or recycle water for on-site use (as in toilet to tap water) to achieve an integrated and intelligent utilization of materials that maximizes their value, prevents pollution, and conserves all resources. Recent reports show that LEED-certified buildings can cut greenhouse gas emissions and water consumption by half while costing 25 percent less to operate and enjoying nearly 30 percent higher occupant satisfaction and lower interest rates. However, LEED "green only" accounts for only 1 percent of the total US building stock.

Green building (sometimes called sustainable or high-performance construction) is the practice of using efficient and sustainable materials during the life of a building from site choice to design, construction, economy, utility, durability, operation, water usage, comfort and convenience, maintenance, renovation, and demolition. There should also be consideration given to nonmonetary benefits such as aesthetics and historic preservation.

Cost, value, and specific needs should be part of ongoing engineering and maintenance programs for the purpose of achieving the required functions and safety at the lowest cost. DART (an interactive tool to guide designers toward Performance-Driven Design) aims to achieve all three values at once.

Performance-Driven Design (PDD) is a strategy to improve the economic, social, and environmental value of the built environment by incorporating flexibility in office design as well as being attentive to psychological aspects.

Building owners, designers, and builders face a future challenge to meet demands for new and renovated facilities that are accessible, secure, healthy, and productive while minimizing any negative impacts on society, the environment, and the economy. Ideally, building designs should result in net-positive benefits.

At present, design advocates are encouraging retrofitting existing buildings rather than building new. Designing major renovations and retrofits for existing buildings to include sustainable design attributes reduces operation costs and environmental impacts, and can increase building resiliency. An existing building should be looked at in terms of both the human labor and material costs that might be saved via renovation as opposed to those that would be incurred in

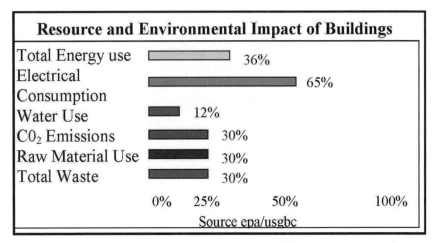

Resource and Environmental Impact of Buildings

Total Energy use   36%
Electrical
Consumption   65%
Water Use   12%
CO_2 Emissions   30%
Raw Material Use   30%
Total Waste   30%

0%   25%   50%   100%

Source epa/usgbc

Figure 10.2.   **Resource and Environmental Impact of Buildings. Source: epa/usgbc.gov.**

constructing a new building in an effort to avoid the squandering of resources that occurs when a building is either demolished or allowed to decay.

DOES SUSTAINABLE (GREEN) DESIGN COST TOO MUCH?

In a survey of over 700 construction professionals, 80 percent cited "higher first costs" as the biggest obstacle to green building—it's the most common criticism of sustainable building, despite the fact that Jerry Yudelson, a green building expert, claims that, "LEED buildings cost 25 percent less to operate and enjoy nearly 30 percent higher occupant satisfaction and lower interest rates."

Buildings.com reports on a post-occupancy evaluation study of twenty-two green federal buildings conducted by the General Services Administration and the Pacific Northwest National Laboratory. The study compared one year of operating data and surveys of green building occupants to those of the national average for commercial buildings. It found green buildings:

- Cost 19 percent less to maintain.
- Use 25 percent less energy and water.
- Release 36 percent fewer carbon dioxide emissions.
- Have a 27 percent higher rate of occupant satisfaction.

Another study found certified commercial green buildings on average cut greenhouse gas emissions from water consumption by 50 percent, reduced solid waste management–related GHG emissions by 48 percent and lowered transportation-related GHG emissions by 5 percent when compared to their traditional California counterparts.

LEED buildings do add about 2 percent higher costs. However, the additional investment typically yields operational savings worth ten times that much. Other studies show that many LEED-designed buildings do not cost more and can actually cost less than conventional construction as they save money in the long run via their sustainable practices.

Numerous sources of funding for green building are available at the national, state, and local levels for industry, government organizations, and nonprofits. To begin, check out https://archive.epa.gov/greenbuilding/web/html/funding.html.

ENVIRONMENTAL QUALITY OF LIFE "INSIDE"

Indoor environmental quality is critical to meet energy, air, and health requirements and to operate a building efficiently. If your building is new, eliminate chlorinated fluorocarbon (CFC-based) refrigerants in heating, ventilation, air conditioning, and refrigeration. Check out:

- Indoor air quality.
- Ongoing energy performance in building.
- Occupant/employee comfort and health in buildings.
- Day lighting strategies for sunshine.
- Distracting noises from outside or from surrounding spaces and occupants.
- Waste management, recycling, waste audits, composting.

OPEN WORKING SPACE

Non-hive-like cubicles and open offices provide greater flexibility and offer workplaces at a lower cost. There are some problems with lighting and noise (60 percent of workers think it interferes with getting their jobs done efficiently).

There are different specifications for office space use measurements, but the average is approximately 151 square feet per worker, according to CoreNet Global. That's down from 176 square feet in 2012 and 225 square feet in 2010.

A feeling of openness is created if your workspace is less than 50 to 75 percent enclosed by walls or windows, has a view to the outside, and does not make workers feel confined. A workplace may feel cramped if a worker is aware of more than eight people around them. Workers should see different perspectives and directions at different times.

SOME SOUND SOLUTIONS

Noises hamper concentration and very different tasks done in close proximity can cause difficulties. Create barriers and sound breaks between sources of noise. Optimize room shape and size to reduce distracting sounds and use acoustic tiles on ceilings and walls to dampen noise, extend walls up, increase insulation partitions, use noise-reduction ceiling tiles, an air-ducted return system, and furniture with 60-inch minimum height that has good sound-absorbing surfaces. Place noise-making office machines in separate rooms away from workers, thereby limiting off-gassing from printers in common work areas as well.

MAINTENANCE

"Commission" is a process in which engineers check and tune up building systems. This can lead to a reduction of 10 to 15 percent in annual energy bills. For a 50,000-square-foot building, this adds up to a whopping $15,000 in savings per year. Commissioning costs from $.05 to $.40 per square foot.

- Green maintenance of buildings means reduced burden on natural resources which means reduced cost.
- Green materials mean easier maintenance. Not only are green products better for health and safety reasons, they are typically more durable and last longer, saving staff time and cutting maintenance and operating costs for the property.
- Healthy materials mean healthy tenants. A building that uses materials that do not emit toxic gases will help children and all building occupants breathe easier.
- The number of landfills is growing at an alarming rate. A building that uses recycled materials and also recycles its construction and demolition waste will divert a lot of waste from going to landfills, which are some of the largest producers of methane gas.

11

Smart Buildings

An Intelligent Choice

Ever been first into the office on a frosty morning and raced to turn on the space heater under the desk, or jacked up the thermostat hoping to get a blast of cool air on a stifling summer afternoon, or discovered that the lights are burned out in the conference room, or realized that no one ordered paper or printer cartridges? These problems and a myriad of others will be taken care of by a new addition to the work world—welcome to the smart building.

A smart building is one that uses several technologies and information systems to optimize total building performance. A building automation system (BAS) is a "smart," centralized nerve center of a building's service procedures. Its purpose is to control, monitor, and optimize building services. This includes lighting, heating, cooling, ventilation and filtration, safety and security, closed-circuit television (CCTV), access control (grants or revokes the right to access some data or perform some action), audiovisual and entertainment systems, time and staff attendance movement and availability (most notably tracking staff and patients in hospitals), and tracking resources, supplies, and maintenance. It does this while being "green" in the process by optimizing and reducing energy consumption and operating costs, monitoring and maintaining electronic devices, and improving utility service, thus lowering the cost and carbon footprint of a business.

FACTS AND FIGURES

- Smart is about gauging and calculating data management because you can't manage what you don't measure.
- Using sensors means more and smarter monitoring.

- Application software leverages the capabilities of a building just like it does for your phone.
- Smart means responding to and serving the buildings needs efficiently.
- Working smarter comes from using data that is collected effectively to achieve productive results.
- Smart means knowing what customers need before they ask for it.
- User-friendly and easy-to-read and interpret, dashboards track, analyze, and report energy performance trends and make technology accessible and workable for users across all skill levels. Potential benefit: up to 3 percent reduction in HVAC energy consumption.
- Operational guidelines help clients maximize insight into their building's compliance with standard operating procedures. The potential benefit is a 4 to 8 percent reduction in HVAC energy consumption.
- Automatic fault detection capabilities provide a prioritized and action-able view into dozens of potential building equipment failures, allow-ing companies the opportunity to reduce the frequency and impact of such conditions as unbalanced airflows, broken dampers, control com-ponent failures or degradation, air-cooled condenser fouling, and valve leakage. The potential benefit is 10 to 15 percent savings in HVAC energy consumption.
- Supervisory control is maximized by correlating and integrating infor-mation from all building equipment to ensure that comfort, temperature, humidity, and ventilation requirements are met in the most energy-effi-cient way. The potential benefit is 8 to 12 percent reduction in HVAC energy consumption. Additional savings of 2 to 5 percent are achievable by automating set-point temperature changes.

TODAY AND TOMORROW

The fact is that today most of our offices are just plain dumb. They are too cold or too hot, equipment is poorly managed and maintained, and resources and energy are wasted. However, intelligent building technologies are gain-ing traction in an ever-growing number of offices. Taking advantage of a smart building can lead to significant financial gain for property owners, and increase production of those working within these structures. Investment in these technologies and building energy management systems (BEMS) in particular makes good business sense, and the outlook is bright for a future of energy-wise, comfortable, and sustainable workspaces.

In the past automated systems to heat, cool, ventilate, and light our build-ings (accounting for about half of the energy demand) were primarily used by manufacturers that offered complex and costly systems designed to meet the needs of only the largest buildings.

In the late 1990s the automation and controls industry heard the rumblings of disruption as Silicon Valley began jumping at the opportunity to control and operate energy efficiency in buildings via software. Consequently, a new market of intelligent building solutions bloomed. Software analytics, or BEMS, are at the core of the intelligent building concept. The hot, lucrative market in global smart building solutions is expected to grow from $2 billion a few years ago to more than $10 billion at present and will be more than $25 billion by 2020, according to a report from IDC Energy Insights.

Although the origins of intelligent buildings and building management systems have roots in the industrial sector in the 1970s when they were used to automate production processes and to advance plant performances, those concepts and applications were then adapted, developed, and automated during the 1980s. Until recent years energy efficiency has been a relatively low priority to building owners and investors. However, with the dramatic increase and awareness of energy use cost and anxiety, and the advances in cost-effective technologies, energy efficiency is fast becoming an essential part of building management and operations strategy.

Think of a building as a machine that balances and reconciles saving energy, maximizing comfort, stability, and security in order to provide and modify a comfortable and healthy ecosystem for everyone's needs and preferences. People want to experience the same kind of comfort and convenience in workplaces that they expect at home. This pressure is influencing commercial building owners and managers to adopt intelligent building technologies and BEMS.

The intelligent control of buildings will also become more "digital" with further services available via the "Internet cloud" in the future that will make smart building management systems financially feasible. A smart building management system can transmit data generated from hundreds of buildings to a single command center where facilities' professionals monitor equipment performance. With today's affordable high-capacity computing, a company can use one smart building management service to monitor and control hundreds of facilities around the world.

It will be possible for customers to reserve time on systems permanently or for specific periods of time using smartphones or other devices. Building automation systems could then use digital appointment calendars to arrange users' workstations in line with their preferences before they arrive, and save energy when a user is not in the office. Add it all up and it appears that buildings will increasingly have the ability to listen to and accommodate the wishes of their occupants, and save resources and money in the bargain.

"Allowing employees to individually control their workspace is a concept that is appealing to many companies clamoring to provide new and interesting perks in an effort to recruit and retain sought-after workers in a competitive technology industry," notes Lindsay Baker, of research and marketing at Building Robotics.

Tighter operating budgets, stricter government guidelines and government control of environmental standards, global trends toward improving indoor air quality standards, energy savings and resource conservation, and increasing public awareness are all important drivers of change in the need for smart buildings.

Energy savings potential from BEMS can range from 15 to 40 percent which represents up to 10 percent of the total energy consumption of the domestic residential and commercial sectors. Some of the savings are very dramatic. A good example is a case cited by the *Wall Street Journal* where Web-based analytics discovered that one of eight apartment buildings on a particular property was using one million more gallons of water annually compared to other buildings on the same property. Ultimately, an ill-fitting rubber seal in the tank that controls flushing was found to be the culprit. Detecting and replacing that toilet part led to better building management, greater efficiency, resource conservation, and a savings of $80,000 per year, far more than what was spent on automation and the trivial cost of replacing a rubber seal.

WHAT'S IT ALL ABOUT?—THE INTERNAL CONTROLS

Controllers and sensors are electronic devices designed to monitor, control, and detect changes, along with collecting other information, and to act on that input to alter or modify energy and resources in a building's environment. BAS controllers have the ability to read and control temperature, humidity, pressure, current flow, airflow, and other essential factors. The Siemens building management system in Vienna can access some 10,000 sensors that provide extremely energy-efficient lighting, room temperature, and ventilation control.

HVAC SYSTEMS

Many people cope with the bother of controlling office temperature by adding desk fans, space heaters, covering up vents, conducting cold and hot thermostat wars with coworkers, or by simply leaving the office. Improper temperature, humidity, ventilation, and indoor air quality can also have significant impacts on productivity and health. This is a big hassle for building managers, employers, and employees because it impacts the performance of workers. But now, an employee can communicate to a commercial building's digital system that they would like to be warmer or cooler by using their computer or smartphone app via an Internet browser and get an almost immediate response.

When we are thermally comfortable we work better, relax and breathe more easily, and focus our attention better. For example, a system can automatically start warming an office in the morning or begin cooling it in the afternoon, before and after workers arrive, which is especially useful for empty rooms that are often being heated and cooled at all times even though people are only using them for small periods of time every day.

While not usually a part of the aesthetics of a building, they are critical to its operations and occupant satisfaction and can be monitored or changed with touch screens and smart controls that can manage everything remotely from a desk.

AIR AND WATER QUALITY

Air handlers (AHs) mix and return outside air so temperature, humidity, and filtration conditioning is needed. This can save money by using less chilled or heated water (not all AHs use chilled/hot water circuits) to enhance energy efficiency while maintaining healthy indoor air quality. Internal air quality (IAQ) control ventilation adjusts the amount of outside air based on measured levels of occupancy.

Variable volume air-handling units (VAV) can change the pressure to the VAV boxes by changing the speed of a fan or blower with a variable frequency drive. Chilled water systems are sometimes used to cool a building's air and equipment utilizing a system of chillers and pumps. Temperature sensors measure the chilled water supply and return lines and chillers are sequenced on and off to chill the needed water supply.

Parameter-based heating control manages protection against freezing or frost and generally involves running heating system pumps and boilers when external temperatures reach a pre-set level. Thermostatic radiator valves sense the temperature in a room and throttle the flow accordingly through the radiator or convector to which they are fitted. Proportional control involves switching equipment on and off automatically to regulate output. Other methods can include thermostats, occupancy-sensing passive infrared sensors (PIRs), and manual user control.

LIGHTING CONTROL

Adding daylight to a building is one way to achieve energy savings and a more natural setting in an office building. Natural daylight harvesting can make people happier, healthier, and more productive. With the reduced need for electric light, a great deal of money can be saved on energy as nearly every commercial building can implement lighting systems designed to be dimmed based on the availability of daylight. In addition, by reducing electric lighting and minimizing solar heat gain, controlled lighting can also reduce a building's air conditioning load. Alternatively, in cold weather, daylight harvesting can aid in maximizing solar heat gain to help meet a building's heating requirements.

Zone lighting is switched on corresponding to use and layout to light a large or small area. Light-level monitoring consists of switching or dimming artificial lighting to maintain a light level measured by a photocell.

Passive infrared (PIR) occupancy sensing is used in areas which are occupied from time to time, and can be applied to indicate whether or not people are present and to switch lights on or off accordingly.

HOW BUILDINGS, CARS, AND GRIDS COMMUNICATE

The question of how a workplace and the power grid will be able to handle large numbers of electric vehicles (EVs) can be integrated into the energy management of a building. In the morning an EV arrives at an office building and is connected to a charging station, but it doesn't have to be fully charged until evening when employees go home. During the day, however, acting as extra storage energy units, the vehicles can be used as buffers to deliver electricity to the building to compensate for the lower energy output of a building's alternative energy system. This is an example of the Internet of Energy (IoE)—vehicles communicating and acting as an extension of an electronic energy grid. This technology can also be utilized for electronic appliances in the building to ease electrical flow in peak times, allowing the grid to "borrow" electricity from appliances during times of need.

When a building management system obtains information from charging stations regarding the requirements of the vehicles and appliances, it uses this input as well as data from climate-management and other control units to generate an energy demand forecast for the next day. This forecast is sent to the grid operator, who estimates a fixed price for a guaranteed amount of electricity. If the building fails to meet its forecast, it may have to pay a penalty. To prevent this from happening, appliances and electric vehicles at the charging stations can be used as supplemental electricity storage or supply units, thus making it possible to keep an entire building's electricity demand stable and cost-effective.

So-called "smart cities" will be energized by power grids that will be able to balance electricity supply and demand. This will start with smart buildings that estimate occupants' energy needs, integrate vehicle batteries and energy needs into their energy forecasts, respond to changing weather conditions, and automatically alter their behavior to maximize their efficiency.

There are many power plants in the United States that only produce electricity for a minimal period of time, called "peaker" plants, in order to prevent grid overloads during peak load times. These plants are very expensive to operate, so there's a great demand for cheaper solutions. Smart buildings will be equipped with a "smart energy box" that can lower energy demand in a targeted manner during peak load phases, easing the strain on the grid, conserving energy, and saving money by as much as 30 percent, while maintaining the proper level of energy and comfort for building occupants.

Designers and builders are now becoming accountable for increasing the efficiency of energy use. They take into consideration how to manage the resource demands of buildings during construction and remodeling, and provide consumption forecasts for the life of the building while reducing negative impacts on human health and the environment. Building occupants can improve the current standards for performance of efficient water use, energy and atmospheric concerns, and indoor environmental quality, and they have the ability and responsibility to help make these changes when they understand the need for conservation on all fronts.

In fact, when human observation and technology were combined to determine which devices should be automated, energy savings rose to 38 percent; however, when workers were not at the controls to offer their input, the savings disappeared, according to a Chinese study at the Zhong Yuan Institute of Technology.

COST AND SAVINGS

The average cost of using a building management system (BMS) to convert to a smart building is around $2.30 per square foot, or $250,000 for 100,000 square feet, according to *Buildings* publication. This makes a building owner's return on investment initially a little thorny.

But with the Internet of Things (IoT), low-powered networks and inexpensive sensors help keep smart building costs down. Adding IoT-based controls and monitoring for a building can cost only $5,000 to $50,000, a small portion of traditional BMS costs.

Commercial buildings could save billions if investments in energy efficiency were ramped up by just 1 to 4 percent, according to a study by the American Council for an Energy-Efficient Economy (ACEEE). According to *RCR Wireless News*, commercial buildings could save up to $60 billion between 2014 and 2030 with comprehensive BEM programs. Benefits gained from adjusting corporate tax legislation to encouraging the replacement of inefficient equipment and removing regulatory barriers to combined heat and power projects could reduce national energy consumption radically and enhance conservation, saving the economy close to $300 billion, according to the ACEEE.

Remember that smart systems are still machines, however sophisticated the hardware and software may be. Only a combination of machines and people can assess and calculate a building's energy needs in a way that will generate better savings and conservation.

12

The Heat Island Effect
Hot in the City and Why

Mother Nature doesn't like it when urban areas replace open land and changes occur in the landscape when buildings, roads, and other built and hard infrastructures replace surfaces that were once porous, moist, and temperate. The land transforms and becomes tough, dry, and hot due to structures, machines, and the building materials used in construction. The consequence of this phenomenon is called the heat island effect (HIE) and it can radically change the temperature and the quality of life for a city's inhabitants.

Figure 12.1.   Infrared photo of inner city heat. Source: nasa.org.

According to the US Environmental Protection Agency (EPA), temperatures in US cities can get as much as 10 degrees (Fahrenheit) higher than the surrounding nonurban areas. These changes cause metropolitan regions to become warmer than their rural surroundings, forming an island of higher temperature called an urban heat island (UHI).

FACTS AND FIGURES

Buildings require enormous amounts of energy and this produces huge amounts of heat. Increasing energy demand generally results in greater emissions of air pollutants and greenhouse gas discharge from power plants. Higher air temperatures also promote the formation of ground-level "bad" ozone that is not emitted directly into the air but is created by chemical reactions between nitrogen oxides (NOx) and volatile organic compounds (VOCs) in the presence of sunlight. Breathing ozone can trigger a variety of health problems, particularly for children, the elderly, and people of all ages who have lung conditions such as asthma. Ground-level ozone can also have harmful effects on sensitive vegetation and ecosystems.

Studies by the Urban Heat Island Pilot Project (UHIPP) found the hottest spots coming from rooftops, with temperatures measuring up to 160°F. Substances such as asphalt retain 95 percent of the sun's energy during the day. A city like Phoenix can suffer a rise in temperature of 10 to 20 degrees in the summer due to the HIE.

The coolest areas are those covered with vegetation and bodies of water. These areas have what is called a low albedo value, referring to the measure of the reflectivity of the Earth's surface. Something that appears white reflects most of the light that hits it and has a *high albedo value;* dark surfaces have a low reflective (albedo) value and absorb heat. Nearly 40 percent of the increase in temperature is due to the prevalence of dark roofs, dark pavement, and the declining presence of vegetation.

On a hot, sunny, summer day, the sun can heat exposed, dry, urban surfaces to temperatures hotter than the air, while shaded or moist surfaces—often in more rural surroundings—remain closer to air temperature. During the night, the principal reason for overheating is that buildings block surface heat from radiating into the relatively cold night sky. Another effect of buildings is the blocking of wind, which also inhibits the cooling of air by convection (hotter and therefore less dense material rises, and colder, denser material sinks, which consequently results in transfer of heat).

Waste heat from power plants, automobiles, air conditioning, industry, and heated exhaust air from buildings further increases temperatures outside and is a secondary contributor to the HIE.

Heat islands can affect communities by increasing summertime peak energy demand via air conditioning costs, air pollution, and greenhouse gas emissions, and can also contribute to general discomfort, respiratory difficulties, heat cramps, heat exhaustion, nonfatal heat stroke, and heat-related mortality. In fact, heat causes more deaths in the United States than any other natural phenomenon.

An increase of a single degree in summer temperatures causes up to a 4 percent increase in peak demand load for air conditioning, suggesting that the demand for electricity used to compensate for the heat island effect is actually exacerbating it.

Hot surfaces transfer their excess heat to storm water by as much as 25 degrees, which then drains into storm sewers and raises water temperatures as it is released into streams, rivers, ponds, and lakes. Rapid temperature changes can be stressful, even fatal to aquatic ecosystems.

The EPA holds free, national, quarterly urban heat island (UHI) Webcasts. Through these Web casts, and conference calls, people from around the country inform each other of their urban heat island–related work and mitigating effects. http://www.epa.gov/heatisld/resources/webcasts.htm

MITIGATING THE HIE

Some of the factors contributing to the heat island effect are simply out of our hands—like weather, climate, geography, and topography. However, we can take a number of common-sense measures to reduce the effects of summertime heat islands.

The HIE can be countered by using white or reflective materials to build houses, roofs, pavements, and roads, thus increasing the overall albedo effect in the city. Replacing or recovering dark roofing requires the least amount of investment for the most immediate return.

Cool roofs can lower air pollution and greenhouse gas emissions and solid waste generation due to less frequent need for re-roofing and less energy demand for cooling. A cool roof made from a reflective material such as vinyl can reflect three-quarters of the sun's rays and 70 percent or more of the solar radiation absorbed by the building's structure. Using light-colored or "cool"concrete has proven effective in reflecting up to 50 percent more light than asphalt, reducing the surrounding temperature. Cool pavements can improve storm water management and water quality, increase surface durability, enhance night time illumination, and reduce noise.

Trees and vegetation lower surface and air temperatures by providing shade and through evapotranspiration (ET), a term used to describe the amount of evaporation and plant transpiration (the process where plants absorb water through the roots and then give off water vapor through pores in their leaves) from the land surface to the atmosphere.

Evapotranspiration alone or in combination with shading can help reduce peak summer temperatures by 2° to 9°F. Shaded surfaces may be 20 to 45 degrees cooler than the peak temperatures of unshaded surfaces. Plants are most useful when planted in strategic locations around buildings, especially shading pavement in parking lots and on streets. Researchers have found that planting deciduous trees or vines to the west is typically most effective for cooling a building and lowering air conditioning demand, especially if they shade windows and part of the building's roof. By reducing energy demand, plants decrease the production of associated air pollution and greenhouse gas emissions. They also store and sequester carbon dioxide, and reduce particulates and urban noise. Vegetation reduces runoff and improves water quality by absorbing and filtering rainwater.Tree shade can slow deterioration of street pavement, decreasing the amount of pollutants created by the production of more asphalt and cement and lessening the need for repaving and maintenance. Trees and vegetation provide needed aesthetic value and habitat for many animals.

The city of Los Angeles estimates that an extra $100 million is spent each year on climate control due to the heat island effect. In a ten-year projection, Los Angeles urban temperatures will drop by approximately 5°F after planting ten million trees, reroofing five million homes, and painting one-quarter of the roads at an estimated cost of $1 billion. This means annual savings estimated at around $170 million in energy, and an additional $360 million in healthcare, with the project paying for itself in only two years.

A five-city study on costs revealed that for every dollar spent, communities received payback ranging from $1.50 to $3.00 in benefits.

URBAN HEAT ISLAND EFFECTS

- Rainfall rates downwind of cities are increased between 48 and 116 percent. One of the results of this warming is the increase of monthly rainfall by about 28 percent—between 20 to 40 miles downwind of cities, compared with upwind.
- HIE increases the occurrence of weak tornadoes.
- Growing seasons in seventy cities in eastern North America were about fifteen days longer in the edges of urban areas compared to rural areas.
- UHI may alter local wind patterns, affect the development of clouds and fog, vary humidity, and contribute to additional rain and thunderstorm activity.

GREEN, COOL ROOFS

Green roofs, also known as living roofs, sky gardens, vegitecture, or agritecture, have been in existence for thousands of years for use in winter insulation, summer cooling, beauty, and edible landscaping. Basically, they are a layer of vegetation grown on a rooftop or on the side of a building, reducing temperatures of surfaces, filtering and channeling water runoff, and mitigating the generally hot environment of a building. Rooftops account for 5 to 35 percent of the total building area of cities, so there's significant potential to help address the problem of UHI.

In many urban areas the built environment has replaced almost all vegetation. Building materials hold hundreds of times more heat than natural landscapes and increase the absorption of heat and the radiation of that heat to surrounding areas.

Green roofs are more effective than conventional roofs for cooling buildings because they utilize a heat-transfer mechanism known as evaporative cooling, which is unavailable in most conventional roofs.

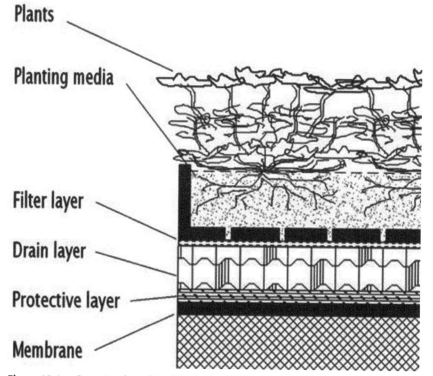

Figure 12.1.   Cross Section of Basic Green Roof Elements. Source: greenroofs.com.

And a tasty bonus is that many roofs present a place to grow produce—a boon to many inner-city neighborhoods, sometimes called "fresh food deserts," where fresh fruits and veggies are hard to find.

BENEFITS OF GREEN ROOFS

- Properly insulated, green roofs can help a building retain up to 80 percent of its indoor climate's heating or cooling.
- Green roofs sometimes supply energy via wind and solar power.
- They can be as simple as 2 inches of groundcover or as complex as a fully accessible park complete with mature trees.
- They create a habitat for wildlife and provide people with an urban oasis.
- Green roofs can garner LEED points that can lead to energy, tax and insurance rebates.
- They can help to lower respiratory distress.
- A green roof helps to ensure the structural integrity of a building.
- It aids passive solar cooling.
- Green roofs maintain the waterproofing of buildings.
- A green roof ensures, fosters, and engineers the long-term health of urban plants.
- It supplies and supports natural biodiversity.
- Green roofs can be installed on a wide range of buildings, from industrial facilities to private residences.

GREEN ROOF COSTS

- The initial costs of green roofs are higher than those for conventional ones. This is offset by reduced energy use, storm water management, longer lifespan, and the added benefits of emotional and physical health to the people and animals of the community.
- Estimated costs of installing a green roof start at $10 per square foot for simpler roofing to $25 per square foot and up for more complex roofs.
- A 21,000-square-foot green roof would cost $464,000 to install versus $335,000 for a conventional roof (in 2006 dollars).
- However, over its lifetime, the green roof would save about $200,000, making it an attractive alternative both in terms of money and for the health of people and their surroundings.

COOL ROOFS

A high thermal emittance (the ability for materials to release stored heat) is important in climates that are warm and sunny. These properties help roofs to absorb less heat and stay up to 50° to 60°F cooler than conventional materials during peak summer weather. Reflective roof coating can often reduce your peak cooling demand by 15 to 20 percent.

Cool roofs absorb some heat gain during the winter and they provide net energy savings, especially in areas where electricity prices are high. A study by the Heat Island Group at Lawrence Berkeley National Laboratory found that in sunny climates, buildings with white roofs required up to 40 percent less energy for cooling than those with black roofs. At current utility rates, that means you could save $120 or more per year in cooling costs. A California study found that cool roofs may provide an average yearly net savings of almost fifty cents per square foot.

Cool roofs generally gain their characteristics through various materials and coatings. Elastomeric coatings have elastic properties and can stretch in the summertime heat, and then return to their original shape without damage. Cool roof coatings on a low-slope roof might cost $0.75 to $1.50 per square foot, while single-ply cool roof membrane costs vary from $1.50 to $3.00 per square foot.

The cost of cool roofing materials versus conventional ranges from zero to 5 or 10 cents per square foot more for most products, or from 10 to 20 cents more for a built-up roof with a cool coating used in place of a smooth asphalt or aluminum coating. Cool roofs recoup this initial added expenditure via decreased cooling costs in summer, savings from downsizing cooling equipment, and their longevity compared to conventional roofs.

A cool roof transfers less heat to the building below, so the building stays cooler and uses less energy for air conditioning. Cool roofs help to prevent heat-related illnesses and deaths.

REFLECTIVE ROOF COATING BENEFITS

Cool roofs can achieve LEED credits and federal tax credits for EnergyStar-qualified roofing materials. Tax credits for solar energy systems are available at 30 percent through December 31, 2019. The credit decreases to 26 percent for tax year 2020, and 22 percent for tax year 2021, and then expires December 31, 2021. For details, please visit EnergyStar's Federal Income Tax Credits for Energy Efficiency page (https://www.energystar.gov/about/federal_tax_credits).

- To get an idea of how much you can save, see the EnergyStar cool roof calculator at http://rsc.ornl.gov/.
- For a list of suitable reflective roof coating products, check out the US Environmental Protection Agency's Web site at https://www.energystar.gov/products/building_products/roof_products.

COOL PAVEMENTS

This term refers to materials that reflect solar energy, enhance water evaporation, mitigate roadway flooding, or have been modified to remain cooler than conventional pavements. It includes a range of techniques and materials used for heat reduction.

Conventional paving materials can reach peak summertime temperatures of 120° to 150°F, transferring excess heat to the atmosphere and heating storm water as it runs off the pavement into local waterways. Due to the large area covered by pavements in urban areas (nearly 30 to 45 percent of land cover), they are an important element to consider in diminishing heat islands.

Cool pavements can be constructed with existing paving technologies (such as asphalt and concrete) as well as with newer approaches, such as the use of coatings or grass paving (turf grown in or around a semi-rigid form).

Cool pavement technologies are not as advanced as other mitigation strategies and there is no official standard or labeling program to designate cool paving materials. However, the Transportation Research Board (TRB) Design and Construction Group has established a Subcommittee on Paving Materials and the Urban Climate to address the influence of pavements in the formation and mitigation of the UHI and to examine the relationship of pavements to broader climate concerns. The subcommittee's scope includes modeling, design practices, testing, standards development, planning, and policy considerations.

BENEFITS INCLUDE

- Lowering the ambient outside temperature.
- Porous pavements can allow storm water to soak into or through the pavement and soil, reducing runoff and filtering pollutants and can improve safety by reducing water spray from moving vehicles and increasing traction through better water drainage.
- The open pores of permeable pavements can reduce tire noise by two to eight decibels and keep noise levels below 75 decibels overall.

- Both permeable and nonpermeable cool pavements can also help lower the temperature of runoff, resulting in less thermal shock to aquatic life.
- Enhancing water quality.
- Reflective pavements can enhance visibility at night, potentially reducing lighting requirements and saving, money, energy, property, and lives.
- Cool pavements in parking lots, on playgrounds where children gather, or in other areas where people congregate provide a more comfortable environment.
- Estimates for the benefits of cool pavements, as part of heat island mitigation, are difficult to calculate based on temperature reduction alone.

13

Sustainable Transportation

People Power and Alternatives to Petroleum Fuel

In suburban settings, the love of cars is ardent, obsessive, and zealous. On roads and highways in the United States, the single-passenger commuter is the rule rather than the exception. And the long lines of moms in their SUVs waiting to pick up a single child is almost comical. By the way, SUVs put out 43 percent more global-warming pollutants (28 pounds of carbon dioxide per gallon of gas consumed) and 47 percent more air pollution than the average car.

A large part of work is getting there. People who commute regularly will average one workweek each year stuck in traffic, or one full year of their lifetime commuting. There are an estimated 128.3 million commuters in the United States. An annual Gallup Work and Education Survey finds that American workers spend from 46 to 90 minutes commuting to and from work on a typical day. Close to one-third of all energy consumed in the United States is for transportation of people and goods, and the infrastructure in need of repair amounts to about $3 trillion.

It seems that the only way to dethrone obsessive drivers from their bucket seats is with a powerful explosive or a very persuasive argument. According to the Internet magazine *RMI Outlet*, "Today's vehicles are overdesigned, underutilized, overloaded with equipment, inefficient, polluting, and—thanks to the drivers behind the wheel—dangerous. And the average personal vehicle sits parked for 90 percent of its life."

When we do drive our cars we tend to drive alone—more than 75 percent of American commuters are solo drivers, even though our vehicles are designed for multiple occupants. Empty third-row seat, anyone?

This leads to so much traffic that we spend 38 extra hours per year sitting in a paved purgatory of our own making. Some people believe that transportation is a basic human need, but gas-powered vehicles are the most expensive way to get around. For long-distance travel, they are pretty good. For short-term, they are a disaster. Boulevards used to be for public use. Then came the streetcars and trolleys and a fast and efficient way to carry many passengers claimed the streets. But it was the auto that finally sounded the death knell of efficient inner-city transportation when we started to build cities around auto accessibility. Automobile companies coined the term "jaywalking" to shame pedestrians running afoul of roads and actually lobbied for jaywalking laws in addition to blaming pedestrians for their own injuries and deaths caused by autos.

Cars take up 100 square feet each. Traffic is becoming gridlocked in most cities and one reason is that public transportation has not been used enough. People complain that they should not pay public subsidies for transportation that they don't use. But this thinking is beginning to change. Public transit ridership in the United States last year hit its highest level since 1956, ironically the year president Dwight Eisenhower signed the Federal Aid Highway Act that created the interstate highway system, fueling a car culture that sapped public transit ridership.

Then there are the peripheral costs of cars, measurable in environmental damage, healthcare expenses, injuries, fatalities, vehicle upkeep, DMV fees, insurance, wasted time, and space for parking. Plus, 95 percent of a car's energy goes toward moving the car itself, and only 5 percent to moving the passenger, and the rate of power lost to the axle is about 30 percent of fuel burned. Some economists and planners have suggested that drivers who choose to use congested roads should actually pay for them, calling it "Congestion Pricing."

Although US cities have largely evolved around the personal vehicle, the average city has more cars than parking spaces and spends a large portion of its budget maintaining roads and other vehicle infrastructure. Unnecessary parking spaces and unnecessary roads can be converted to needed living spaces and pedestrian malls, parks, bike paths, and additional business space. Most city car lots take up a great deal of available space simply to store vehicles when they are not being used. Parking lots are city deserts. They don't employ many people and they blight cities, displacing neighborhoods and emptying pocketbooks with monthly parking rates in some large cities creeping up past $500 a month and hourly rates at $6 to $10 per hour, close to the hourly wage of lower-end laborers in some cities.

One fix is to just build more roads to reduce traffic and make more space for vehicles, but that only exacerbates the problem. Called "induced demand," it happens when we build more roads, especially in cities or heavily trafficked areas that just fill up with more cars. In other words, the amount of traffic goes up by the same amount as the roads built to carry it—so we get nothing more than added traffic and the ills that accompany it.

Another thing—cars are unsafe. Many accidents are caused by people who think they are excellent drivers. It seems that overconfidence and inattention are primary causes of a majority of accidents, since most car crashes happen on sunny days in good weather. Modern cars with techno-gadgets can make the situation worse by lulling drivers into thinking they are safe, instigating more overconfidence—which, again, causes still more accidents.

Owning a car is expensive. It is estimated to cost $0.61 per mile to operate a personal vehicle or an average of $10,000 a year, according to the American Automobile Association. This translates to a large percentage of yearly income for many people. Frequently, more money is spent on vehicles than for food, making annual car expenses second only to housing for many consumers; and some people think it's more important to own a vehicle than to have a place to live. And it has fattened one of the enormous industries in the world—as insurance is mandatory state law for vehicle operators.

Each year US drivers spend well over $1 trillion on auto expenses, including about 2 billion barrels of petro-gas and oil, producing about 1.5 billion tons of carbon dioxide—one-quarter of all US emissions annually. Traffic congestion is a primary reason that dramatically affects the emitting of pollutants, including nitrogen oxide, sulfur dioxide, carbon dioxide, and ozone. These toxins can have severe impacts on public health, especially in urban areas, and contribute to global climate change. In addition to fouling air quality, traffic congestion causes economic impacts, including delays, an inability to accurately predict travel times, increased fuel consumption, and potential safety hazards.

The indirect societal cost of these vehicles, including pollution, lost productivity (sitting in traffic among other things), land use for roads and parking lots, road construction and maintenance, plus injuries and fatalities, that cost us $2 trillion per year, brings the annual total to a staggering $3 trillion, according to *RMI Outlet* (a site whose mission is to drive the efficient and restorative use of resources).

If all the cars on the planet could get 60 mpg, instead of 39 mpg, it would remove 1 billion metric tons of carbon from the atmosphere, the same as eliminating 800 coal-fired power plants.

CARPOOLING

One of the last of the great indulgences is the single-car commuter. More than ever, people are driving alone to work, regardless of fuel prices. In the last decade, the number of people driving alone to work increased slightly to 77 percent; only 8 percent drive with someone else, a whopping 80 percent claim that they aren't interested in carpooling, and 60 percent of commuters claim they like the time spent alone in their cars, according to a US Census Bureau report. In addition, carpooling use dropped and the share of commuters using public transportation stayed the same.

A considerable amount of saving can be had by carpooling. Share your ride and the gas bill with just one friend and you each save $650 a year. Fill the car with additional commuters and you each save nearly $1,000 annually, excluding parking fees, some sharing of minor maintenance costs, and tax deductions.

FACTS AND FIGURES ON GETTING THERE

- Nine billion gallons of fuel are wasted in traffic each year—800 times the amount of oil spilled by Exxon Valdez.
- The average American commutes to work 16 miles each way, and the average car gets around 23 mpg, which equates to about 7 gallons of gas per week to commute.
- Close to 143 million Americans aged 16 and older commute to work each day. That's about 45 percent of the population that's on the move at any given time.
- Commute times haven't changed much over the past decade or so. It takes workers about 25.4 minutes on average to get where they're going.
- More than 109 million commuters make their daily trip solo, which is about 76 percent of all commuting workers total.
- Only about 13 million workers carpool, and 10 million of those ride with just one other person. Around 1.8 million travel in groups of three while about 1.3 million carpool with four people or more.
- 93 percent call driving more convenient than other forms of commuting.
- 88 percent say they'd oppose highway tolls.
- 250 million cars in America are estimated to travel 4 billion miles in a day, according to the DOT, and use more than 200 million gallons of gasoline.
- If every car carried just one more person, we'd save 8 billion gallons of gas a year.
- Traffic congestion wastes 3 billion gallons of gas a year.
- Proper inflation of tires can improve gas mileage by more than 3 percent. Every increase in fuel efficiency makes a difference as every gallon of gasoline saved keeps 20 pounds of carbon dioxide out of the atmosphere.

Many city planners are looking to creating walkable cities like those in Europe and Japan. Things are beginning to change even in a car-chummy city such as Los Angeles. The city is removing space from streets and giving it back to people. The Los Angeles Department of Transportation is planning subways and rail lines and improving bus lines. More people are walking and biking every day, and some bridges and overpasses are being reclaimed as public parks.

Companies like the tech-savvy cab company Uber, city-wide share cars (like Zipcars), cheap rental cars, and Go-Cars (small cart-like vehicles), self-driving vehicles, delivery drones, and Mobility on Demand (MoD) transportation projects are becoming innovative models of mobility in the inner cities. Pickup trucks shouldn't be used to deliver a bouquet of

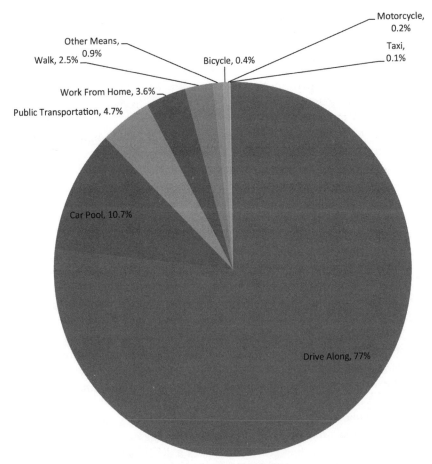

Figure 13.1.   How People Get to Work. By Jeff Dondero.

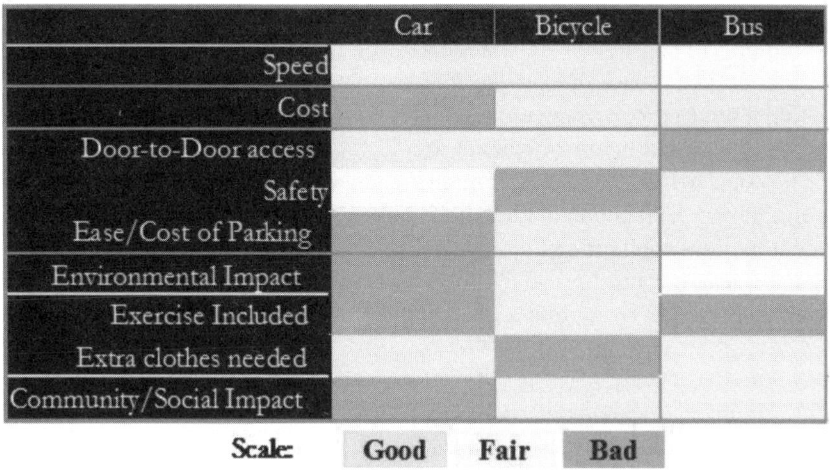

	Car	Bicycle	Bus
Speed	Good	Bad	Good
Cost	Bad	Fair	Good
Door-to-Door access	Good	Good	Bad
Safety	Good	Fair	Good
Ease/Cost of Parking	Bad	Good	Good
Environmental Impact	Bad	Good	Good
Exercise Included	Bad	Good	Fair
Extra clothes needed	Good	Fair	Good
Community/Social Impact	Bad	Good	Good

Scale: Good Fair Bad

Figure 13.2. Comparison Choices for Short Trips. Source: dot.gov.

flowers or a huge SUV for solo personal errands. Not since the horse and buggy gave way to Ford's Model T has mobility been in the midst of such a radical change.

It will take more than one lifetime to totally rid us of the petro-powered engine and our transportation infrastructure of diesel trains and trucks, planes, ships, and the refineries and gas stations that serve them. So the obvious answer right now is to increase efficiency and conservation, use public transportation, and plan for a petro-free future.

TAKE A WALK

Just about all the alternatives to driving are healthy. Some of the reasons people can't walk to work are distance, time, and weather. But when you can walk you should. A 140-pound person walking at 3 mph will burn approximately 80 calories per mile; this converts to getting roughly 360 miles per gallon of gas. Walking helps with mental health by increasing the levels of mood-lifting chemicals in the brain. It can provide a distraction from daily worries and an opportunity for increasing social contact by exercising with other people.

On a daily basis, each additional hour spent driving is associated with a 6 percent increase in the likelihood of obesity, while each additional mile walked is associated with a 5 percent reduction of that likelihood.

We also know that some 47 percent of people with cardiac problems show symptoms of depression and that depression is as much a risk factor as smoking, poor diet, or lack of exercise. Regular physical activity leads to a 17 to 28 percent reduced risk of developing depression. Researchers have found those who walk regularly are 24 percent less likely to develop dementia.

Physical activity regulates hormones such as estrogen, testosterone, and insulin. It may help people with Type 2 diabetes in their overall health management by lowering blood glucose levels as muscles use more energy. Walking also affects the speed by which food passes through the bowel, reducing the chance of cancer by around 25 percent.

BETTER TO BIKE

To be nitpicky, even bicycling uses fossil fuels, if you consider what goes into producing the bicycle and the food to fuel the cyclist, but it's still 75 percent more efficient than a car. Bicycling is 117 percent more efficient than walking and using a bicycle to commute four days a week for four miles (one-way) saves 54 gallons of gas annually. Half of the average trips that people make are three to five miles or less, considered a perfect distance for a bike ride. Bicycling can reduce mortality by 35 to 40 percent, and 40 percent of Americans say they would commute by bike if safe parking facilities were available.

If we spent our gas money on food per average portion to fuel biking, that money would take us 26 miles on beef, 48 miles on potatoes, 106 miles on beans, and 109 miles on rice. And according to *Car-Free World Newsletter*, if you took the money you'd save by getting rid of your car when young and invested it (with a little luck and no recession), you could have $2.3 million by the time you retired.

The average price for a car in the United States is $33,560. A good bicycle sells for approximately $300, costs almost nothing to operate, maintenance is low, and it lasts for years. The average annual operating cost of a bicycle is around $300, about 2 to 4 percent of the yearly cost of $13,000 to $15,000 for an average car. The energy and resources needed to build one medium-sized car could produce 100 bicycles.

A 150-pound person will burn about 500 calories riding a bike at a leisurely pace of 30 minutes to work and 30 minutes home. At that rate, you'll shed 5 to 10 pounds in about two or three months. Biking or walking to work eliminates the need to exercise at the gym and just three hours of riding per week can slash your risk of heart disease and stroke in half, and you'll have better looking legs, too. On the same urban route, car drivers were exposed to more airborne pollution than cyclists, despite the cyclists' higher respiration rates. Over 95 percent of the muscle energy we use at the pedals is translated into forward motion and less than 5 percent is lost.

Call the Bicycle Federation of America at 202-332-6986 to see what kinds of bicycle programs your city has and get more information on how to encourage cities to develop more bicycle paths and alternative transport incentives.

PUBLIC TRANSIT

Six in ten Americans have public transit available, but only 10 percent use it regularly, even though public transit is viewed as 170 times safer than automobile travel, according to the National Safety Council. A bus with as few as seven passengers is more fuel efficient per gallon than a single person in a vehicle used for commuting. With all fifty-five bus seats filled, it equates to 330 passenger mpg; with 70 percent filled, the efficiency would be 231 passenger mpg. The use of public transportation can drop per mile costs to $0.15, unlocking an annual savings to the country of $1 trillion.

With more than 480,000 school buses on the road each day, that's nearly 17.3 million fewer vehicles on the streets, saving an estimated 2.3 billion gallons of fuel each year as well as reducing congestion, GHG emissions, and road wear and tear. A single school bus can eliminate approximately thirty-six cars. And most buses have accommodations for carrying bicycles.

Taking public transit saves an average household $6,000 to $9,000 on automobile expenses per year and compared to a car public transportation produces 95 percent less carbon monoxide, 92 percent fewer volatile organic compounds, 45 percent less carbon dioxide, and 48 percent less nitrogen oxide per mile. Research shows that by using public transit, the typical automobile driver can reduce individual daily carbon emissions by 20 pounds or more.

Public transportation saves 1,500 million gallons in fuel consumption annually. That's 150 times the 10 million gallons spilled in Alaska by the Exxon Valdez. For every 10,000 solo commuters who leave their cars at home and use public transportation service for one year, it reduces the national fuel consumption by 2.7 million gallons.

A new study released by the American Public Transportation Association (APTA) shows that a person can reduce his or her chance of being in an accident by more than 90 percent simply by taking public transit as opposed to commuting by car.

PLANES

Air travel produces large amounts of emissions, so reducing how often you fly by even one or two trips a year can reduce air pollution significantly. Long-haul flights produce, on average, twice as much in *emissions* per mile traveled per passenger than cars, and short-haul flights produce three times as much and emit a toxic stew of other harmful gases like nitric oxide and nitrogen dioxide, water vapor, soot and sulfate particles, sulfur oxides, and carbon monoxide in addition to carbon dioxide.

Gases released in the upper atmosphere where planes cruise have a much greater impact than gases released on the ground due to something called the "radiative forcing" effect (RF is the measurement of the capacity of a gas or other forcing agents to affect that energy balance between incoming solar radiation and outgoing infrared radiation, thereby contributing to global heating and climate change).

It takes only one trip to Europe from the west coast, a trip to Hawaii from the east coast, or a couple of cross-country flights to do as much damage as an entire year of commuting and cruising in a car. Leaving a fluorescent light switched on for nine months straight produces the same amount of GHG.

SOME SUGGESTIONS FOR TRANSIT HELP

- Check out the Federal Transit Administration for subsidies: www.fta.dot .gov/funding/grants/grants_financing_3732.html
- Department of Transportation: www.ehow.com/list_7469781_carpool -rideshare-grants.html
- Many transit agencies are listed on the Internet at: www.fta.dot.gov or http://www.apta.com/
- Talk to your company about setting up transportation and carpooling plans for employees.
- Many companies have ride-sharing programs as do many local governments. Call the American Council on Transportation at 1-800-223-8774 for ride-sharing information in your area.
- Talk with fellow employees and employers about carpooling and creating alternative work schedule programs such as flextime and telecommuting. These programs shift work hours away from peak commute times, decreasing traffic congestion, commuting time, driver stress, and miles on your car.

14

Alternative Energy Choices

Our future will be determined by the choices we make today. Some choices are chancy, others inevitable, but what they bring about is yet to be seen. Perhaps the biggest decision is between polluting and nonpolluting, sustainable and nonsustainable forms of energy, and how we're going to make changes.

Simply said, nonsustainable forms of energy are those that pollute, like petroleum products, that will not always be readily available and can be exhausted. Sustainable forms of energy are nonpolluting and don't emit excess carbon dioxide, greenhouse gases, and other particulates into the atmosphere, are not contributing to amplified climate change, and are self-sustaining, meaning they will always be available.

As the facts and fantasies about various biofuels are investigated, it becomes clear that there are many eco-benefits to be had by replacing petroleum fuels with bio-alternatives. Since biofuels are mostly derived from agricultural crops, they are inherently renewable—our farmers can produce them domestically, reducing our dependence on unstable foreign sources of oil and easing the depletion of our national treasure. Additionally, biodiesel and ethanol emit less particulate pollution than traditional petroleum-based gasoline and diesel fuels. Biofuels only put back into the environment the CO_2 that the source plants absorbed out of the atmosphere in the first place—or what's called zero net CO_2 gain.

There is no quick-fix for weaning ourselves off of petroleum, but some of the alternatives, such as natural gas, are part of the strategy for reducing dependence on petro-fuel. Alternative fuels such as natural gas, ethanol, and methanol—while useful as a bridge from fossil fuels—still emit greenhouse gases and soot particulates, a byproduct of natural gas that darkens snow, trapping heat and causing glaciers to melt faster.

The future will likely see a combination of energy sources from biofuel, flex fuel, and fuel cells to hybrids and all-electric cars. A flexible-fuel vehicle or dual-fuel vehicle uses an internal combustion engine designed to run on more than one fuel, usually gasoline blended with either ethanol or methanol fuel. Hybrids use a combo of fuel and electric power, and fuel cells are powered by a hydrogen-run electric vehicle motor and discharge only water vapor.

Businesses or individuals may be eligible for an income tax credit of up to 50 percent of the equipment and labor costs for converting vehicles to operate using alternative fuels. Qualified alternative fuels are comprised of compressed and liquefied natural gas, propane, hydrogen, electricity, and fuels containing at least 85 percent ethanol or methanol. The maximum credit is $500 for the conversion of vehicles with a gross vehicle weight rating (GVWR) of 10,000 pounds or less, and $1,000 for vehicles with a GVWR of more than 10,000 pounds.

SUSTAINABLE FUELS

Here's a rundown on the pros and cons of the most probable and common choices of power sources to run power plants, buildings, homes, and vehicles.

- Biomass energy is made from plants and plant-derived materials, with cellulose (plant matter) being the largest resource used.
- Geothermal energy uses the heat from the Earth via steam reservoirs, geothermal reservoirs, and tapping heat from shallow ground near the Earth's surface.
- Hydrogen is a clean-burning fuel that may be combined with oxygen in a fuel cell, which produces heat and electricity for EVs.
- Hydropower, or hydroelectric power, occurs when the energy of moving water is captured and turned into electricity.
- Ocean energy uses heat energy stored in the Earth's oceans to generate electricity through a process called ocean thermal energy conversion (OTEC) or the energy of tidal action.
- Solar energy uses the sun as a powerful source of energy, as from solar panels that produce either passive or active (photovoltaic) power.
- Wind (eolic) energy uses a wind turbine to generate electricity.

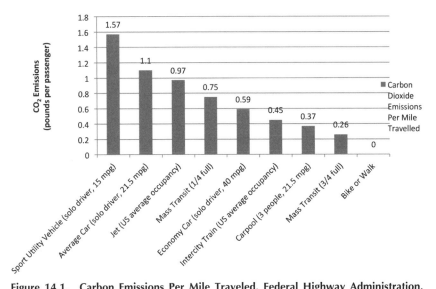

Figure 14.1. Carbon Emissions Per Mile Traveled. Federal Highway Administration. Source: www.fhwa.dot.gov.

SUSTAINABLE POWER TO RUN VEHICLES

Biodiesel is a form of fuel manufactured from vegetable oils, animal fats, insects, fish, weeds, or recycled restaurant greases via a process called transesterification. The process creates two products: methyl esters (the chemical name for biodiesel) and glycerin (a valuable byproduct usually used in soaps and other products).

Biodiesel can be used alone or blended with petro-diesel in various proportions and can also be used as heating oil.

Pros

Biodiesel is renewable, widely available, and older diesel cars can burn it. Used vegetable oil is sometimes free and can be domestically produced from renewable sources. Biodiesel fuel actually produces approximately 25 percent more power than gasoline for the same volume (depending on the mixture) and creates 93 percent more energy than is needed to produce it.

It's the only alternative fuel to have fully completed the health effects testing requirements of the Clean Air Act and is nontoxic, biodegradable, and safe to handle. Biodiesel is significantly better than ethanol in terms of environmental impact. It reduces harmful pollution by an average of 40 to 45 percent over regular diesel. It creates fewer air pollutants (other than nitrogen oxides, which are gaseous air pollutants produced by the combustion of fossil fuels and contribute to the formation and modification of other air pollutants), and reduces particulates and unburned hydrocarbons, which reduces health risks.

Biodiesel can be used in its pure form, but may require certain engine modifications to avoid maintenance and performance problems. One hundred percent biodiesel is referred to as B100, 5 percent biodiesel is B5, 2 percent biodiesel is labeled B2.

Biodiesel is a better lubricant than petro-diesel so it helps to extend the working life of engines. It is easier to ignite than petro-diesel, meaning more complete, efficient combustion, hence less wear on an engine.

Blending B100 with petroleum diesel may be accomplished by mixing in tanks at the manufacturing point, splash mixing in the tanker truck, or in-line mixing the two components.

In the United States, soybeans are the main source of biomass for this biofuel, but in other countries plants like rapeseed are used and there is also a large potential for biodiesel derived from biomass algae.

Because biofuels are produced locally, manufacturing plants can employ hundreds or thousands of domestic workers, creating new jobs, and reducing the energy needed to distribute it.

Cons

If biodiesel were to be used on a wider scale, more plants for oil would be grown as cash crops, reducing the amount of food grown to feed people and animals in less economically developed countries.

The quality of biodiesel varies widely at present so carmakers will only honor warranties of vehicles using up to 5 percent biodiesel. Biodiesel costs more than petroleum diesel and, so far, supply issues have prevented biodiesel from becoming readily available. A common misconception is that a vehicle will run on used vegetable oil or grease right out of the vat, but there is a small refinement or cleaning process required.

Biodiesel can congeal (become gel-like) in colder weather and can clog filters due to the release of deposits that attach to tank and pipe walls from previous diesel use. This problem is most associated with mixes over B20, however once the old deposits have been discharged, the clogging of filters no longer becomes an issue.

One of the problems with the fuel itself is the increase in NOx emissions (nitrogen oxides are a close relative of dioxides) from the reaction between nitrogen, oxygen, and hydrocarbons during combustion, especially at high temperatures.

The EPA found that the use of B20 can reduce fuel efficiency by 1 to 2 percent. Many refineries have been unable to produce biodiesel that meets American Society for Testing and Materials (ASTM) quality due their inability to remove all impurities and water during the washing and refining processes.

B100 and other diesel and biodiesel blends are more expensive to consumers than the standard #2 diesel. This is the result of the rapidly rising feedstock prices and production problems. B100 is generally not suitable for use in low temperatures. According to the EPA, B100 can cost anywhere from $1.95 to $3.00 per gallon at present, while B20 blends average about 30 to 40 cents more per gallon than standard diesel. This all depends on variables such as the raw material used and market conditions. There are currently around 1,000 filling stations in the United States that carry some blend of biodiesel, the majority of which are in the Midwest due to the local availability of natural resources and biodiesel plants.

On average, there is about a 10 percent reduction in power with biodiesel. In other words, it takes about 1.1 gallons of biodiesel to equal 1 gallon of standard diesel. Even though a regular diesel engine can run biodiesel, that does not mean that it is ideal.

Massive quantities of water are required for proper irrigation of biofuel crops as well as to manufacture the fuel, which could strain local and regional water resources.

ETHANOL

Ethanol is one of man's oldest and favorite synthetic substances, dating back 9,000 years, and is basically high-powered alcohol. Ethanol was first used as a fuel in 1876, but because ethanol is an alcohol it was banned by Prohibition in 1919, gaining favor again for a short while in 1974 during the gas shortage. At present it is being used to blend with gasoline, although the big bang-for-the-buck promise of ethanol has largely been debunked. At present this alternative fuel still seems to be one way to help reduce pain at the pump and decrease the release of GHGs into the atmosphere.

Ethanol fuel mixtures have "E" numbers which describe the percentage of ethanol in the mixture by volume. E85 is 85 percent anhydrous (free of water) ethanol and 15 percent gasoline. Low-ethanol blends, from E5 to E25, are also known as gasohol. The most common use of the term refers to the E10 blend. Use of blends of E10 or less is led by the United States, where ethanol recently represents 10 percent of the gasoline fuel supply.

Pros

Ethanol is a renewable fuel that comes from agricultural feedstocks and plants that contain a large amount of sugar or components that can be converted into sugar, then ethanol, such as sugar beets and sugar cane, corn, wheat, and barley, which can stress food and feed sources and prices. It can be produced domestically and is considered renewable, reducing demand for foreign oil imports.

Scientists are experimenting with other forms of vegetation that are weeds and don't replace food sources while lowering GHGs emissions. Sawgrass is a cellulosic material of which 100 percent of the plant is utilized, as opposed to corn that uses only the protein kernel. It produces ten times more energy than corn, does not need petroleum-powered fertilizer, can be cultivated on land considered non-farmable, uses fewer if any herbicides, and emits fewer GHGs than corn.

Existing cars can use E10 blends and more than 8 million cars already on the road can use E85. One Btu of fossil energy consumed in producing and delivering corn ethanol results in 1.3 Btu of usable energy in your fuel tank. Ethanol is cheaper than gas, but not by much, and costs about the same as ethanol, propane, and natural gas.

The amount of GHGs reduced by ethanol ranges from 19 to 52 percent, according to the DOE. Ethanol displaces the use of toxic gasoline components such as benzene, a carcinogen. Seventy-four percent of ethanol is broken down and is nontoxic within five days, making spills far less worrisome than for petroleum products.

Ethanol fuel can work in a standard gasoline engine without any adjustment. Most automobiles available in the United States are flex-fuel capable and there are roughly 2,000 stations already serving E85. While most of these stations are lumped in the Midwest, they are increasing nationwide.

Cons

Ethanol production consumes close to 6 billion bushels of grain, or about 40 percent of the corn produced in the United States, according to the USDA Economic Research Service. It is expensive because it uses a lot of land, water, and energy and impacts corn prices, which have tripled in the last thirty years. It takes 450 pounds of corn to make one tankful of gasoline for an SUV, or enough protein to feed one person for one year. Producing enough ethanol to replace America's imported oil alone would require putting nearly 900 million acres under cultivation—or roughly 95 percent of the active farmland in the country, let alone other manufacturing needs, some of which are petroleum-based.

A gallon of E85 usually costs about the same as a gallon of regular gasoline, but doesn't offer as much energy and decreases gas mileage by 20 to 30 percent.

Although there are about 170,000 filling stations nationwide, less than 1 percent of US gas stations carry E85 and it is only widely available in the Midwest. Other areas, even populous ones, have little E85 infrastructure. And filling up with ethanol on the West Coast or in the mid-Atlantic states may cost a driver as much as 40 cents more per gallon than gas.

Ethanol absorbs water and is corrosive, which makes it difficult to ship through existing pipelines to other regions of the United States from the Midwest where most production occurs.

METHANOL

Following the crude oil price shocks of the 1970s, methanol blending of 3 to 5 percent in gasoline was first introduced in the early 1980s. Today the blend is 85 percent methanol and 15 percent gasoline and is called M85.

Methanol, or methyl alcohol, is also known as wood alcohol and is a clean-burning, high-octane blended component made from alternative nonpetroleum energy sources. Methanol can be made from a wide array of feedstocks (any renewable, biological material that can be used as a fuel source).

Currently, virtually all methanol produced in the United States uses methane as a byproduct derived from natural gas. However, methane also can be obtained from fossil fuels, fermenting organic matter (biogas), methane hydrates (ice-like combinations of methane and water on the sea floor), the passed gas from cows, and the byproducts of sewage and manure.

Methanol contains only 67 percent of the energy of gasoline per gallon and takes twice as much methanol to produce the same amount of power, but it's far cheaper to produce. A recent report from the M.I.T. Energy Initiative, "The Future of Natural Gas," called methanol, "the liquid fuel that is most efficiently and inexpensively produced from natural gas."

Pros

Methanol is a potent fuel with an octane rating of 100 that allows for higher compression and greater efficiency than gasoline. To companies with rich deposits of oil and natural gas, it is highly advantageous for them to make it into methanol and use it as a gasoline extender.

M85 vehicles are flex-fuel users meaning any mixture of M85 and gasoline in the fuel tank can be used by the engine. It's cheaper to produce methanol fuel than ethanol and as gas prices increase, methanol becomes a much better alternative and can easily be converted into hydrogen, offering a promising future for use in fuel cells.

Methanol is biodegradable and will dissolve in contact with water and is a great deal safer than gas and ethanol due to the fact that it's far less flammable and can be used to generate electricity to charge EVs and hybrids.

Cons

Methanol is not volatile enough to start a cold engine easily. It attracts water, making it corrosive to some materials and requires special materials for storage and delivery.

Methanol has only 51 percent of the Btu content of gasoline by volume, which means the miles per gallon ratio is worse than gas and ethanol. The lower energy content and the higher cost ratio in building methanol refineries compared to those for ethanol distilleries have relegated M85 to the backseat.

Over a twenty-year period, methane is estimated to have a warming effect on Earth's atmosphere 84 times that of carbon dioxide, as methanol produced from natural gas results in a net increase of CO_2. It is far more toxic than ethanol and creates a high amount of formaldehyde in emissions.

However, with the dramatic increase in drilling and fracking, the lower carbon dioxide emissions from natural gas compared to other fossil fuels' methane emissions have come under increased scrutiny; especially following the Aliso Canyon disaster in California which released an estimated 97,100 metric tons of methane into the atmosphere. By that measure, the Aliso Canyon leak produced the same amount of GHGs as running 1,735,404 cars for a full year.

HYDROGEN FUEL CELLS

The fuel cell is not really a new technique for producing power. It was first pioneered by the scientist Christian Friedrich Schönbein in 1838. William Grove, a chemist, physicist, and lawyer, is generally credited with inventing the fuel cell in 1839.

Hydrogen (H) is the most basic and most common and abundant element in the universe. It's the gas that combines with oxygen to make water (H_2O), and with carbon to form fossil fuel compounds. Hydrogen isn't actually an energy source, but it carries the energy that's created when it's produced.

Think of a fuel cell as a battery that you add fuel to, in order to keep it running. It can use various types of fuel and is the cleanest energy source since the only byproduct is distilled water (when using hydrogen as a fuel source). If fossil fuels are used as the original source of hydrogen, there will be by-products such as carbon dioxide.

The key to using hydrogen as a energy source is the fuel cell consisting of an electrolyte (a substance that produces an electrically conducting solution) sandwiched between two electrodes (a conductor through which electricity enters or leaves an object) via connectors (basically plus and minus anodes and cathodes) for collecting the current that converts the hydrogen energy into electricity. Although it's the most environmentally friendly method of creating power, at present it's also very costly.

Hydrogen does not occur freely in nature in useful quantities, but it is manufactured in a number of ways. It can be made from natural gas or it can be made by passing electric current through water. The fuel cell concept can be used to provide electricity to rural areas with no power lines, in very small things like mobile phones, in very large industrial generators for businesses used by companies such as Staples, Walmart, Coca-Cola, FedEx, and various factories, and most recently in automobiles.

Fuel cells are perfect companions for renewable energy sources like solar or wind, because when the sun goes down or the wind stops, the fuel cell can take over for totally continuous and clean energy. They can run nonstop as long as fuel is available, have no moving parts, and create no noise or pollutants when fueled by hydrogen.

Almost 90 to 95 percent of the energy generated by fuel cells is transformed into electrical energy, which makes the process highly energy-efficient. Hydrogen is twice as efficient as gasoline, can be refueled just as quickly as gasoline, and hydrogen fuel cell–vehicles can get up to 480 miles on one tank.

According to the Fuel Cell and Hydrogen Energy Association (FCHEA), stationary fuel cell systems, which are commercially available, reduce the staggering costs and inconvenience of power outages, offer insulation against physical and cyber-attacks, voltage surges, and/or power variations. Stationary fuel cells provide constant, clean power generation and create less than one ounce of pollution per 1,000 kW-hours of electricity at 80 percent efficiency, compared to 25 pounds of pollutants created by more traditional combustion systems for the same amount of electricity.

Fuel cell versions of the electric car can be refueled on-the-go in less than 10 minutes. They can be equipped with other advanced technologies to increase efficiency, such as regenerative braking systems, which capture the energy during braking and store it in a battery. Several foreign and domestic companies are manufacturing fuel cell–powered cars and making "bank" that they will become popular in the near future.

Pros

Fuel economy is about twice that of gasoline vehicles. Hydrogen is abundant and can be made from renewable energy sources. Hydrogen handling has had an excellent safety record. Because hydrogen gas is so light, if a fuel tank were punctured or otherwise compromised, the hydrogen gas would instantaneously dissipate into the atmosphere.

Cons

An acceptable range for vehicles requires extremely-high-pressure, on-board hydrogen storage, and there are few places to refuel. Hydrogen is very expensive to transport and there is no infrastructure in place yet. Currently hydrogen fuel is made from nonrenewable natural gas in a process that creates enormous CO_2 emissions, and is very costly.

For proper performance, hydrogen-powered cars have some temperature restrictions. At places where the temperature goes below freezing, there are chances that the water in the fuel cells may freeze. In case of damage, the cost for the repairs and replacement of the fuel cells is also expensive, and the maintenance of these cars is expensive. Performance as far as speed and power are disappointing at present.

Even though hydrogen is an abundant element, after it has been refined for fuel, it can get between fifty-five to eighty miles per gallon. And although at present hydrogen runs between $2.50 and $3 a gallon, it is still a better buy than other fuels.

At this time the technology is vastly underdeveloped and faces serious challenges that could keep it off the roads and out of your driveway for some time.

NONSUSTAINABLE FUELS

In 1859, Edwin Drake and E. B. Bowditch of the Seneca Oil Company drilled the first commercial oil well in the United States in Titusville, Pennsylvania. The well produced less than ten barrels a day. Today the United States is the second largest producer of petroleum in the world. One oil barrel (bbl) contains approximately 42 gallons of crude oil, from which a US refinery can generally refine roughly 19 gallons of gasoline.

Gasoline is truly a compromise between convenience and functionality. And like it or not, petroleum products are the fuel of most American transportation systems.

Petro-fuel will be around for a long time—just about anything you use and touch—including the manufacture of wind turbine blades and solar panels—

What Do We Pay for in a Gallon of Gasoline

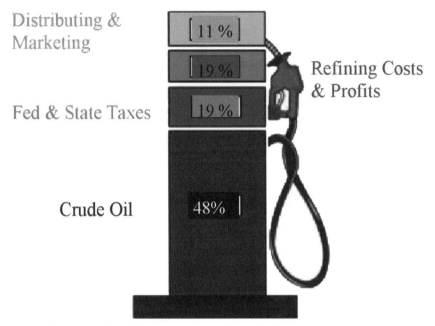

Distributing & Marketing

Fed & State Taxes

Crude Oil

Refining Costs & Profits

[11 %]

19 %

19 %

48%

Figure 14.2. What Do We Pay for in a Gallon of Gas?

has something to do with fossil fuels. Replacing America's 255 million cars (with an average lifespan of more than 11 years) with non- or low-emission vehicles will take decades.

Table 14. 1. CO_2 Emissions from US Electric Power Sector

Source	Million Metric Tons	Share of Total
Coal	1,364	71%
Natural gas	530	28%
Petroleum	24	1%
Other	7	<1%
Total	1,925	100%

Source: useia.gov.

Pros

Fossil fuels are cheap and reliable sources of energy, technologies and availability of petroleum have been well developed, the cost of extraction of petroleum is relatively low, and it requires reduced physical and technical effort. The distribution system for transportation is vast, and it can be easily transported and stored.

Petroleum products can move vehicles longer and faster as compared to other types of energy sources. It is abundant and easy to use. Diesel engines get 30 percent better fuel economy, contain more useable energy than gasoline, deliver more power for their size, and any diesel car can run on a blend of renewable biodiesel fuel.

Diesel engines are very long-lasting (some engines can clock up to a million miles), get better mileage than gas engines, and are relatively easy to maintain. Diesel engines release hundreds of millions of pounds *fewer* carbon dioxide emissions than gasoline engines.

Cons

Processing and extracting oil contributes to pollution and environmental degradation. Emissions contain huge amounts of carbon dioxide, contributing to climate change and also to health dangers.

Petro-fuels are nonreplenishable and nonsustainable. Oil spills are a gross violation of the environment costing billions of dollars and causing irreparable damage to the land, water, and animals. Every year, the United States alone is responsible for 1.7 billion tons of CO_2 released into the atmosphere from the tailpipes of gas-powered cars, according to the EPA. The sixteen largest petro-fueled ships in the world produce more pollution than all the cars on Earth.

Gasoline engines burn fuel inefficiently, as only 20 to 30 percent of potential energy is converted into turning wheels for a standard internal combustion vehicle.

Prices are propped up by government tax and production incentives (almost $5 billion in 2014. Solar received a little over $1 billion). Every year, the average driver can incur around $3,000 a year in gasoline expenses, and this amount is unstable due to fluctuating fuel costs.

The EPA estimates more than 57 percent of the oil Americans use comes from foreign sources that are hostile to the United States. The Organization of the Petroleum Exporting Countries (OPEC) manipulates the cost of oil, giving it the power to affect our economy.

NATURAL GAS

When most people think of natural gas, they think furnaces, dryers, and ovens, even barbeques because, comparatively speaking, there aren't many compressed natural gas (CNG) vehicles in the US. At present the federal government offers a tax credit of $4,000 to buyers of CNG-powered vehicles to help curb air pollution. States may offer additional credits for both the vehicle and a home fueling device.

Pros

Natural gas costs less than gasoline, is the cleanest transportation fuel available today, and provides comparable power. The United States has one of the largest reserves, enough to last at least 100 years.

Natural gas vehicle (NGV) engines have a very high compression ratio, burning most of the fuel and leaving very few byproducts behind, and according to the EPA, NGVs can reduce carbon monoxide emissions by 90 to 97 percent and nitrogen oxide emissions by 35 to 60 percent when compared with gasoline. They can also potentially reduce non-methane hydrocarbon emissions by 50 to 75 percent.

A regular gasoline-powered car averages 32 miles per gallon while a compressed natural gas or CNG-powered car averages 43 miles per gallon. Even if you cannot afford to buy a natural gas vehicle, you are still able to morph your car into a hybrid with a conversion kit making it a flex-fuel vehicle.

CNG is safer than gasoline and is a clean-burning fuel that will not explode even if subjected to a round of rifle shots.

Cons

Despite what the petro-fuel companies say, natural gas is still a pollutant compared to sustainable fuels. It emits about sixty times more air pollution and greenhouse gas per unit of energy generated than alternative sustainable fuels.

Huge gas tanks are needed in vehicles, reducing trunk space, and carry the equivalent of only a few gallons of gasoline. CNG provides limited range, and there are few places for consumers to refuel. In most of the country, refueling is relatively slow, and the range between fill-ups is about 220 to 250 miles.

There are about 1,000 natural gas fueling stations across the United States, but only 536 are available for public use, according to the US DOE's Alternative Fuel Data Center Web site (www.afdc.energy.gov)—the majority of which are clustered around major metropolitan hubs.

Natural gas creates about ten times the amount of bird kills per unit of energy from air pollution, ground degradation, and habitat destruction.

Converting a car to CNG can be costly with prices ranging from one to several thousands of dollars. Natural gas vehicles typically cost $25,000 to $75,000 more than similar diesel- or gasoline-powered vehicles. Although

Congress encourages CNG conversion by offering tax cuts of up to 50 percent, the conversion still needs an EPA certification to qualify for the tax credit. Getting a certificate can be costly and time consuming.

Natural gas vehicle acceleration is typically slower than gas. Maintenance issues require some special training for the technicians who service them.

PROPANE

Propane is found mixed with natural gas and petroleum deposits and is one of the many fossil fuels that are included in the liquefied petroleum (LP) gas family. Most people are familiar with propane from backyard barbecues and fuel for remote houses.

Pros

Propane-powered vehicles have the longest driving range of any alternative fuel—more than 250 percent farther than CNG, about 60 percent farther than methanol, and 25 percent farther than ethanol blends.

Propane yields about 12 percent less carbon dioxide and 20 percent less nitrous oxide emissions than gasoline and leaves no carbon build-up inside your engine. After 5,000 miles, your engine oil will still look new.

It has a higher octane rating than gas, between 100 to 112, and is generally cheaper than gasoline, it's reasonably available, and burns much cleaner than gasoline. Propane is easy to transport and store.

Virtually any internal combustion engine can be converted to propane from a car engine to a lawnmower engine and will run well at sea level or at high altitudes. Conversion kits for a V8 engine will run about $1,000 to $1,200. Tax credits and grants are available to help offset the higher purchase price for natural gas-powered vehicles.

Cons

Propane needs far more room than the conventional gas tank to carry enough fuel to make it worthwhile.

Propane makes about only about 74 percent of the power of gasoline. One gallon of propane is equal to about nine-tenths of a gallon of gas and availability may be a problem.

If you live in extremely cold environments (10°F or below) you may need an engine block heater. Maintenance issues require some special training for the technicians who service them.

A WORD ABOUT FRACKING:

To extract natural gas, a technique called fracking has been employed, which has caused a lot of controversy in communities where it is mined, in the media, and in the courts recently. Fracking is the technique of injecting water, sand, and chemicals at high pressures into shale and other tight rock formations to release the fuel trapped inside. Combined with horizontal drilling it has allowed us to access huge amounts of heretofore unrecoverable natural gas.

Alarmingly high methane emissions come from oil- and gas-producing fields due to fracking. If methane—a potent greenhouse gas—is leaking, it could be offsetting much of the climate benefit of the ongoing shift from coal to gas-fired plants for the generation of electricity.

Methane released during fracking doesn't just end up in the air, but also in the water. Fracking fluid that is injected into the wells contains a toxic soup of hundreds of chemicals, including carcinogens and volatile organic compounds like benzene, toluene, ethylbenzene, and xylene. And thanks to heavyweight lobbying, fracking is exempt from the Safe Drinking Water Act and the companies that hide behind this act do not have to reveal exactly what their fracking brews contain. Consequently, that makes it tough for the EPA is regulate fracking, toxic disposal, and water discharge.

Fracking can take anywhere from 12 to 20 million gallons of water for a single well. Additionally, drilling operations have caused contamination of surface and drinking water. Wells are fracked as many as eighteen times and most of the water injected underground is not recoverable; wastewater can contain huge amounts of brine (salts), toxic metals, and radioactivity and is unfit for further use, usually requiring disposal in an underground injection well.

In addition, a large percentage of the water, about 70 percent, is currently reused, but that still leaves a massive amount of toxic wastewater to be disposed of. Some companies re-inject the wastewater back into the ground. Unfortunately, this method has the inconvenient habit of causing an earthquake every now and again—making Oklahoma, a large producer of fracked gas, the earthquake capital of the United States.

Another alternative is waste treatment by first removing the contaminants and then dumping the so-called clean treated water into a nearby sewer. But standard municipal water treatment plants cannot handle the level of contamination, especially radioactivity, found in these waters.

The Environmental Protection Agency has been doing a double-take around the problems of fracking. One side states that natural gas plays a key role in our nation's clean energy future, and that the United States has vast reserves of natural gas that are commercially viable as a result of advances in horizontal drilling and hydraulic fracturing technologies enabling greater access to gas in shale formations. Responsible development of America's shale gas resources offers important economic and energy security benefits.

The EPA is working with states and other key stakeholders to help ensure that natural gas extraction does not come at the expense of public health and the environment. Its newest report produced significant findings; key among them was that fracking has caused contamination to drinking water resources and that "We found scientific evidence of impacts to drinking water resources at each stage of the hydraulic fracturing water cycle," stated Tom Burke, the EPA Deputy Administrator.

Despite heavy lobbying that insists that fracking does not pose a threat to drinking water, now the oil and gas industry will have to defend the controversial drilling technique.

15

Carbon Offsets

Proactive Process or Permission to Pollute?

WHAT ARE CARBON OFFSETS?

Carbon offsets were designed to be a practical and effective way to address climate change and encourage the growth of renewable energy. Using energy or products manufactured with fossil fuels generates carbon dioxide and other greenhouse gas emissions, commonly known as a "carbon footprint," which contribute to the speed of climate change.

Carbon offsets supposedly counteract carbon emissions and contribute to a more sustainable future by encouraging one party to a carbon offset to invest in green resources and alternative energy sources, in hopes of reducing carbon emissions and greenhouse gases.

There are many ways to reduce your personal carbon footprint—by driving a hybrid car, reducing energy use, buying EnergySmart appliances, using alternative forms of energy, employing efficient lighting, eating locally grown food, and watching what you buy and how much you travel. Companies buy carbon offset certificates to mitigate their emissions output and that "offset" money is used to help industry, especially in underdeveloped countries, invest in greener alternatives and energy production. Tit for fat tat, so to speak.

Carbon offsets are controversial for a number of reasons. Many consider it a shell game that allows some companies or individuals to simply absolve themselves of guilt by buying their way out of responsibility for their carbon footprint "sins."

Carbon offsets are not inherently a bad idea under certain circumstances, but is a method that invites corruption, the bad manners of petro-belching, and a diversion from conservation. Amy Moas, Senior Forest Campaigner at Greenpeace, wrote in the California newspaper, *Capitol Weekly*, "In other words, hey, I'm rich, I can indulge myself or my company because I can pay for it. If protecting trees in Mexico so that some companies can pollute more sounds dubious, that's because it is."

Others see it as a way of neutralizing greenhouse gas emissions and assisting businesses and countries that have difficulty trying to move toward cleaner industries and practices. So, countries with high emissions levels can purchase carbon credits from nations that have less of a negative impact on the environment, but are "industry" poor. The reasoning for this is to make an effort to offset the quantity of greenhouse gases released into the air and to aid underdeveloped countries or companies to become green-clean by providing them with funds to employ nonpolluting practices.

Carbon offsets can be as small as a plane ride, or as far-reaching as plans for future factories. Some businesses, or even organizations of countries, can create a carbon pool for their enterprises. Many of the projects that carbon offsets support are in and of themselves "good" projects deserving of our support because they are worthwhile, not just because of our clean carbon culpability.

The idea for carbon credits came from the Kyoto Protocol of 1997, a United Nations Framework Convention on Climate Change, regarding the prevention of greenhouse gases. It placed a monetary value on the cost of polluting the air. A "credit" is a measure representing one metric ton (a mass equal to 1,000 kilograms or 2200 pounds) of carbon dioxide emissions.

The credits or "offsets" are established in a market where carbon credits are listed, sold, or exchanged between national and international businesses. At current levels, credit for carbon "tax" emissions in the United States is $30 per ton.

These credits are supposed to be used to finance carbon reduction schemes between trading partners. Carbon credits are typically generated by projects which produce renewable energy—through wind, solar, hydro, or renewable biomass—which take the place of fossil fuel power generation.

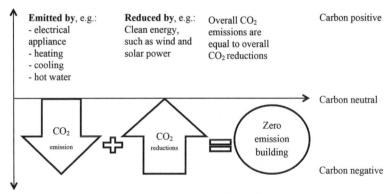

Figure 15.1. Reducing CO_2 Emissions. By Jeff Dondero.

CO$_2$ FACTS AND FIGURES

Average CO$_2$ Emissions Produced per Year:

- Average global citizen—4.5 tons.
- The average US citizen—21 tons.
- Average electricity for one US household—6.2 tons.
- The average US car—4.5 tons.
- 500 megawatt gas-operated power plant—1.5 million tons.
- Coal-fired power plant—8.3 million tons.
- United States as a whole—6 billion tons.
- The planet Earth—more than 25 billion tons (the United States comprises about one-fifth of the world population, but produces about one-quarter of CO$_2$ emissions).

This Causes a Ton of CO$_2$ Emissions:

- Traveling 2,000 miles on a plane.
- Driving 1,500 miles in a large SUV.
- Driving 1,900 miles in a mid-sized car.
- Driving 6,000 miles in a hybrid auto.
- Running an average household for 60 days.
- Using a computer for 10,600 hours.

This Offsets 1,000 Tons of CO$_2$:

- Changing from a large SUV to a hybrid for one year.
- Running a 600kW wind turbine for a year.
- Replacing 500 100-watt light bulbs for 18-watt CFL bulbs.
- Replacing 2,000 old refrigerators with new, high-efficiency models.
- Installing 125 home solar panels.
- Planting an acre of Douglas fir trees.
- Protecting four acres of rain forest.

WILD WEST?

Like a new gold rush, some carbon offset providers are setting up shop by the dozens, ready to relieve consumers of their guilt and cash. But with little regulation, the industry is coming under increased scrutiny. It has been called "the Wild West" by insiders because, for an industry selling a very intangible product, it's often difficult to know exactly what's for sale, how to go about getting it, or if there are any guarantees of what good it is supposed to accomplish.

Perhaps there is reason for the lack of transparency. Profit-based companies account for 43.4 percent of the money raised from their offsetting, which actually goes toward alternative projects, while the nonprofits' average is 81.6 percent.

HOW MUCH FOR A TON OF BUSINESS AS USUAL?

Offsetters also charge very different amounts. According to a review of the industry by Clean Air-Cool Planet (a nonprofit company promoting clean sustainable solutions), offsets range anywhere from $5 to $25 per ton, with the average at $10 per ton. While the reviewers state that there is no link between price and quality they warn, "If something sounds too good to be true, it probably is."

Shell Oil purchased 500,000 carbon offset credits in 2013 from a forest project outside of California in order to allow its refinery in Martinez, California, to keep emitting large amounts of CO_2. For each of those credits, Shell's refinery released an additional ton of carbon dioxide into the air. If this kind of workaround becomes business as usual, companies like Shell will have little reason to invest in cleaner technologies and can dodge the problem of the hidden costs not factored into offsets, like increased health problems in local communities that don't reap the benefits of offsetting.

The other side of the coin is an investment—and it can be as little as $5—that might help build wind turbines, capture methane from a landfill, distribute solar cookers in a developing country, aid in land protection or reforestation, and invest in experimental alternative energy projects and conservation ventures. It's possible that offsets aid in achieving positive goals, but does that make up for dirty deeds?

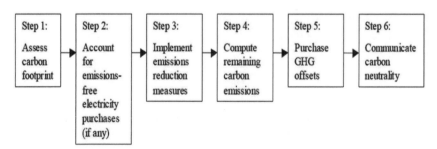

Figure 15.2.   Steps to Purchasing Carbon Offsets. Source: www.yourhome.gov.au.

CARBON OFFSETS

Pros

- In cases where emissions are inevitable, offsets provide a way to do something to mitigate the effects of GHGs.
- The funds received from the sale of carbon credits essentially balances out the higher cost of renewable energy production, making these projects cost-competitive. Without the sale of carbon credits, these projects would not exist.
- Carbon offsets encourage individuals to take responsibility for their carbon footprint and can be considered a positive step towards reducing carbon emissions.
- Currently the highest certification for voluntary carbon offsetting standards is from the Gold Standard, a Swiss-based nonprofit foundation. To be certified, projects must adhere to the standard of using renewable energy or energy-efficient technologies and promote sustainable development.
- Offsets can be a catalyst for change in developing countries where renewable energy projects funded by more industrialized countries could be the basis for new, green, and sustainable growth.
- Offset projects include planting trees to help absorb the excessive carbon dioxide in the atmosphere and to aid in preserving biodiversity, reducing soil erosion and flooding, and providing shade to reduce the need for expensive and energy consuming air conditioning.
- Carbon offsets are beneficial to corporations and people who need to travel a great deal in their businesses.

Cons

Offsets are not a precise science and their use is still relatively new. There is no exact regulation of the credits that people buy and the alternative methods used to offset them. Hundreds of companies are offering everything from kelp beds to wind farms and tree-planting as offsetting options.

Many of the offsetting companies don't have an accurate system to deal with precise and quantifiable carbon measurements for calculating the exact amount of emissions created.

Some organizations and systems of offsetting are considered to be much more accurate than others. Planting trees is generally seen as less valuable than creating more wind and solar technologies because trees only sequester carbon until they die, at which point the carbon will be rereleased back into the atmosphere. Offsets exploit guilt over GHG emissions and underemphasize the culpability for climate change.

Carbon credits do not reduce emissions, and reduction is the main goal. Offsets are likely to be largely ineffective in removing large amounts of carbon dioxide from the atmosphere because once a ton of CO_2 is put into the atmosphere, there's nothing offsetting that can be done to stop it changing our climate. Offsetting accounts for a fraction of all emissions produced; the need is to radically cut our emissions rather than mitigate them.

Critics have also raised concerns over fairness based on the argument that carbon offsetting enables developed nations to continue to perpetrate unsustainable lifestyles by funding carbon projects in developing countries, stating that offsets are also a fictitious commodity created to take advantage of people's concerns over climate change.

The unregulated industry causes problems such as verification, making it difficult for buyers to assess the true value of carbon credits. There are widespread instances of people and organizations using worthless credits that do not reduce carbon emissions or aid in creating alternative energies.

Table 15.1. Weighing Carbon Credit Providers

Evaluative Criterion	Weighting
Providers' Prioritization of Offset Quality	10
Buyers' Ability to Transparently Evaluate Offset Quality	9.4
Transparency in Provider Operations and Offset Selection	9.2
Provider's Understanding of the Technical Aspects of Offset Quality	9.0
Priority Assigned by Provider to Educating Consumers about Global Warming and Global Warming Policy	7.8
Ancillary Environmental and Sustainable Development Benefits of Offset Portfolios	5.6
Use of Third-Party Project Protocols and Certification	3.9

Source: https://www.climatetrust.org/

EXAMPLES OF OFFSETS AND QUALITY RATING

Almost any area or industry can generate high-end or low-quality offsets (e.g., landfill methane, energy efficiency, renewable energy, and reforestation), so it's essential to review a project's characteristics. Examples of both ends of the spectrum include:

Highest Quality: Capturing, flaring (burning off), and verifying methane amounts at a closed coalmine or landfill.

Medium Quality: Paying for the installation of more efficient light bulbs or motors to reduce energy consumption and GHG emissions.

Low Quality: Paying people not to clear-cut trees they had no intention of felling. There have also been cases of groups collecting money for offset projects that don't exist, so an offset project should be *verified* and *enforceable*.

THE GOLD STANDARD

The most rigorous model is the Gold Standard, which was initiated by the World Wildlife Fund (WWF), the *Semantic Sensor Network* (SSN), and HE-LIO International—and endorsed by forty-two nongovernmental organizations (NGOs) worldwide.

Gold Standard certifies developments that support, "A fundamental change in behavior, namely shifting from a fossil fuel-dependent lifestyle and based economy, and designed to ensure that carbon credits are not only real and verifiable but that they make measurable contributions to sustainable development worldwide." Gold Standard certification must include an approved renewable energy supply end use, an energy efficiency improvement project, and reduced production of one of the three eligible greenhouse gases: CO_2, methane (CH_4) and nitrous oxide (N_2O).

They support those offsets that fall into the area of renewable energy: solar, biomass, biogas and liquid biofuels (if they produce electricity), wind power, geothermal energy, small hydro (less than 15 mW).

CARBON CATALOGS

Carbon Catalog, an independent service, is a Web site with a directory of carbon offsets, listing and rating 100 offset providers and 394 projects worldwide. The listings and ratings follow transparent guidelines. Carbon Catalog does not sell offsets or have commercial relationships with providers. Look for the Green-e Climate certified logo on offsets being sold in the voluntary market. Offset Consumer offers a carbon calculator for those wishing to know what their GHG emission tonnage amounts to.

Some responsible and reliable offset guides:

- The Regional Greenhouse Gas Initiative in the Northeast and California's cap-and-trade program include approved offset protocols that meet the specific requirements of each program, under which compliance carbon offsets are issued.
- The David Suzuki Foundation and the Pembina Institute have prepared a *Purchasing Carbon Offsets* guide to help consumers, businesses, and

Table 15.2. Notable Carbon Calculators

	Airplane Travel Emissions	
Atmosfair	https://www.atmosfair.de/index.php?id=5&L=3	Location to location detail, with layovers.
Climate Care	http://www.climatecare.org	Location to location, as well as house and car emissions.
Offsetters	http://www.carbonify.com/carbon-calculator.htm	Location to location detail.
	Business Emissions	
Climate Friendly	http://www.climatefriendly.com/business.php	One of very few business calculators. Includes factory and office electricity, fleet fuel, and corporate air travel.
	Car Travel Emissions	
Certified Clean Car	http://www.icao.int/environmental-protection/CarbonOffset/Pages/default.aspx	Input exact car make and model.
Target Neutral	https://www.targetneutral.com/TONIC/carbon.do?method=init	Calculate up to 4 cars at once.
TerraPass	http://www.terrapass.com/road/carboncalc.php	Input exact car make and model.
Clean Air Pass	https://www.cleanairpass.com/treecanada	Input exact car make and model.
	Other Emissions	
Carbon Counter	http://www.climatefriendly.com/business.php	Calculate "estimated" or "exact" emissions based on your information.
Sustainable Travel International	http://www.sustainabletravelinternational.org/offset/index.php?p=hotel	Includes hotel emissions.
World Land Trust	http://www.carbonbalanced.org/personal/calculator/calctravel.asp	Includes hotel, boat, flight emissions for multiple travelers.
Atmos Clear	http://www.atmosclear.org/calculator_tran.php	Includes household and recreational equipment, from leaf blowers to jet skis.

Source: www.epa.ie

organizations assess the quality of carbon offsets and the vendors that sell them. It includes a survey of twenty carbon offset vendors from Canada and around the world to help shed light on how these vendors are performing.

- The *ENDS Guide to Carbon Offsets 2009* includes a comprehensive listing and rating of over 170 offset providers.
- The *Responsible Purchasing Guide: Carbon Offset*, published by the Responsible Purchasing Network and Carbonfund, is a general overview of how to conduct an inventory and purchase offsets. It includes links to RFPs for the solicitation of bids for the purchase of offset credits.
- Australian Carbon Offset Watch.
- *Carbon Offset Watch Assessment Report by the Institute for Sustainable Futures, University of Technology.*

RESPONSIBLE PURCHASING GUIDE FOR CARBON OFFSETS

- Verified Carbon Standard (VCS) Agriculture, Forestry, and Other Land Use (AFOLU) program; standards established by the Climate, Community and Biodiversity Alliance (CCBA); the CarbonFix Standard (CFS); and the Plan Vivo Standard provide a comparison of the four leading forestry standards in voluntary carbon market and the state of climate forestation projects.
- *Voluntary Carbon Markets: An International Business Guide to What They Are and How They Work* is a general book on the voluntary market and does not include ratings of offset providers.
- AgCert/DrivingGreen (Ireland)
- Atmosfair (Germany)
- Carbonfund (USA)
- Carbon Counter Project
- CarbonNeutral (UK)
- ClimateCare (UK)
- Climate Friendly (Australia)
- Climate Trust (USA)
- co2balance (UK)
- Myclimate (Swiss; USA)
- NativeEnergy (USA)
- Offsetters (USA)
- TerraPass Sustainable (USA)

Do some research and investigation on your own to decide if and from whom you choose to buy offset credits.

QUESTIONS TO ASK YOURSELF:

- What steps have you taken to reduce your own emissions? If the opportunity to go carbon neutral by spending a few dollars online becomes an excuse to not think about what else you can do at home or elsewhere, or lets you feel that it is acceptable to emit more emissions than you might otherwise, then buying offsets may have a negative result.
- In choosing a retail offsets provider, have you paid attention to the quality of the offsets you are purchasing, so that you can credibly claim that you are carbon neutral?
- Is going carbon neutral the beginning of your global warming mitigation journey, or the end? The opportunity to go carbon neutral at an individual level should not become an excuse to avoid thinking about the larger problem of global warming policy. Addressing global warming will require much more than individuals and businesses going carbon neutral.
- What are you doing to leverage your efforts to go carbon neutral? Rewarding with your dollars companies offering carbon neutral products and services? Using your carbon neutrality as a platform to push for global warming policy by your elected representatives? Without public policy, individuals' carbon neutrality cannot solve the problem. Indeed, a key contribution of the retail offsets market may be to promote public understanding and ultimately public policy.

QUESTIONS TO ASK A PROVIDER:

- Do your offsets result from specific projects?
- Do you use an objective standard to ensure the additionality and quality of the offsets you sell? How do you demonstrate that the projects in your portfolio would not have happened without the GHG offset market?
- Have your offsets been validated against a third-party standard by a credible source?
- Do you sell offsets that will actually accrue in the future? If so, how long into the future, and can you explain why you need to "forward sell" the offsets? Can you demonstrate that your offsets are not sold to multiple buyers?
- What are you doing to educate your buyers about global warming and the need for global warming policy?

16

Product Use and Waste

Life Recycle

This chapter is not just about polishing a green apple for the boss, turning off the lights, putting recycle bins in the office, using public transportation or carpooling, drinking tap water, using mugs not plastic cups, using recycled products; the most important thing to recycle is in your head. It has to do with changing your mindset, introducing new habits, and making a conscious effort to conserve by practicing all kinds of resource management.

Chances are that your company is already doing its part to conserve. If not, it's a good thing you bought this book, which means you are ready for change which will better the "bottom line" for the workplace, the worker, and the Earth.

FACTS AND FIGURES

Automation has brought a lot of changes, some of which were unexpected. People thought that computers and e-mail would decrease the use of paper—ironically, the reverse happened. The manufacture and use of commercial paper has increased 245 percent from 1960 to 1994 and continues to grow. Along with the computer revolution and mountains of paper, the throwaway society took hold.

Remember when phones and office equipment lasted forever? According to Wirefly, the average cell phone gets replaced every eighteen months, which adds up to around 100 to 140 million cell phones in the trash every year, representing more than 65,000 tons of e-waste.

The amount of e-waste is becoming humongous. This has raised havoc in our municipal dumps, not only because of the amount of discards, but due to

Figure 16.1. Recycle for Life. By Jeff Dondero.

the toxicity of e-waste. With some simple solutions, this toxic trash can be cut down to size and safely removed, reducing pollution and increasing recycling.

About 85 percent of office paper is currently discarded (over 7 million tons in the United States every year) and is the largest part of office garbage. In the average workplace about 80 to 90 percent of solid waste (mostly paper) is actually recyclable, according to the EPA.

More than four hundred paper mills in the United States use at least some recovered materials in their manufacturing processes, and more than two hundred of those mills use recovered fiber exclusively. Manufacturing recycled paper generates 74 percent less air pollution and uses 50 percent less water than creating paper from scratch. It saves trees, water, and oil, and reduces emissions compared to traditional virgin fiber processes. It saves enough energy to power the average American home for five months. By increasing double-sided copying, offices could reduce annual paper use by at least 20 percent.

OMNIPRESENT PAPER

The two main actions to save paper use are to set up all printers for double-sided (i.e., duplex) printing. The next step is more technical but is worth exploring if you employ a staff of more than fifty. It is called *"pull"* or *"follow-me printing."* This software program reduces printing by at least 30 percent as it solves the problem of finding the closest printer to a worker. The reason it works so well is that over 30 percent of print jobs sent to a printer are never collected and are tossed out.

DISSING DISCS AND DISCARDING CARTRIDGES

Almost 4 million computer diskettes are thrown away every day, which equals over one-and-a-half billion disks per year; stacked, that would be as tall as the Willis Tower in Chicago. Seventy percent of used printer cartridges throughout the world are currently being thrown out and more than 350 million cartridges per year are discarded in US landfills, a rate that is increasing by 12 percent annually. It will take nearly five hundred years for the disks to degrade.

The plastic in each toner cartridge takes almost a gallon of oil to produce. In only twelve months, refilling cartridges instead of manufacturing new ones could save more than one million gallons of oil and save consumers as much as 40 to 50 percent off the price of a new cartridge. New cartridges can cost from $15 to $40, refilled cartridges cost from $7 to $15 and some retailers will refill your cartridges while you shop. In fact, some printers are actually cheaper than the cartridges they use. Digg's tech page claims it is cheaper to buy a new inexpensive printer for the free ink cartridges included in the box instead of buying brand name replacement cartridges that can be very expensive.

An average business that uses one hundred toner cartridge refills in a year creates a carbon offset of one metric ton of carbon emissions. If one year's worth of the world's discarded cartridges were stacked end-to-end, they would circle the Earth more than three times. Every remanufactured cartridge saves nearly 3.5 pounds of solid waste from being deposited in landfills. Many office supply stores offer recycle credit for cartridges.

E-WASTE

Electronic products that have become nonworking or unwanted create the largest stream of discards in the United States and represent more than 3 billion units of office equipment each year, creating 70 percent of overall toxic waste—80 percent of which goes to other countries.

E-Waste is continually on the rise, with an average of 220 tons of computers and other e-waste dumped in landfills and incinerators every year in the United States. Many of these electronics are not designed to be recycled, are difficult to disassemble, and many of the materials simply cannot be recycled. By 2017, the global volume of discarded e-waste will weigh almost as much as two hundred Empire State Buildings, according to a report gathered by the United Nations.

Last year, the US electronics recycling industry processed 3 to 4 million tons of used and end-of-life electronics equipment. More than 70 percent of the collected equipment is manufactured into specification grade commodities—including scrap, steel, aluminum, copper, lead, circuit boards, plastics, glass, and precious metals. However, only about 13 percent of all e-waste is recycled. For every one million cell phones that are recycled, 35,274 pounds of copper, 772 pounds of silver, 75 pounds of gold, and 33 pounds of palladium can be recovered. Recycling one million laptops saves the energy equivalent of the electricity used by 3,657 homes in a year.

Use an e-waste recycler certified through the Basel Action Network (BAN), a nonprofit organization devoted to certifying e-Steward recyclers who are committed to safely and responsibly recycling electronics. The National Technology Recycling Project is a constantly updated, nationwide directory to find the nonprofit computer recycler closest to you. Many manufacturers of electronics have taken the lead to take back their old equipment and/or be involved in the reuse and recycling of their equipment.

According to the 2006 Universal Waste law, it is illegal for residents and small businesses to dispose of fluorescent lamps, household batteries, and other "Universal Waste" in the trash. Electronic devices, such as televisions and computer monitors, computers, printers, VCRs, cell phones, telephones, radios, and microwave ovens, often contain heavy metals like lead, cadmium, copper, and chromium. For a complete list and more details, please visit the Department of Toxic Substances Control Web site.

If you run a business, you're legally responsible for making sure your waste is disposed of properly. The US Department of Agriculture (USDA) offers grants and guaranteed loans that are also available for investments in renewable energy and energy-efficiency improvements for agricultural and small businesses.

PRACTICAL SUGGESTIONS FOR THE OFFICE

- Companies need to create an environmental policy. Encourage and support employee contributions to recycling and emission reduction—they'll come up with a lot of ideas.
- Recycling bins need to be labeled clearly to prevent cross-contamination of paper, metal, and plastic.
- Make sure that recycling bins are located close to printers and photocopiers. People will usually throw paper in the nearest garbage can.
- Explain to the janitorial staff about your recycling initiative, and reach out to building management or the building's landlord to notify them as well.
- Reduce office trash as much as possible and compost and recycle the rest.
- Check your e-waste depot to make sure that the retired office equipment is being recycled properly.
- Reducing the heating or cooling thermostat by one or two degrees can make a huge difference to the office environment and the bills.
- A report by the World Wildlife Fund showed that a majority of companies believe video conferencing and other technologies could reduce dependence on business travel. Also consider a carbon-offset program for business travel.
- Decrease the use of paper products and make sure your paper is nonchlorinated because chlorine is one of the biggest polluters in the papermaking process.
- Set up a bulletin board for memos instead of sending a copy to each employee. Use e-mail instead of paper correspondence.
- Encourage everyone to print documents with soy-based inks, which are less toxic.
- Designate someone to be the "switcher," or ask maintenance or cleaning personnel to shut down a list of designated equipment each evening.
- How the company vehicles are managed and operated will impact the consumption of petro-products and reduce maintenance costs.
- Replace incandescent light bulbs with compact models, and properly dispose of fluorescent light bulbs containing mercury and batteries, which are both highly toxic. Many local and state laws now prohibit discarding these lamps and batteries in trash cans or dumpsters and many retail outlets will take batteries for recycling.
- Use alternatives to polystyrene or Styrofoam products.
- Employ low-flow or timed fixtures encouraging employees to turn off the water while washing hands. Create water-saving landscaping, reuse wastewater, or revamp water-intensive processes. This will also reduce sewer costs by up to 30 percent.

- Use environmentally friendly cleaning supplies.
- Question your suppliers about their environmental policies, and hire vendors that use less packaging material and green shipping policies.
- Weatherize as best you can, or make suggestions to the building owner or superintendent. Check caulking and weatherizing for all windows. Install door sweeps or draft-inhibiting "snakes."
- Use native, local plants, especially those that require the least amount of water.
- Look into LEED, which stands for Leadership in Energy and Environmental Design, for conservation ideas. It offers levels of certification depending on how well a building is engineered with regard to energy efficiency. This can result in an increase in property value and can make a company eligible for significant tax credits from the local, state, and federal government.

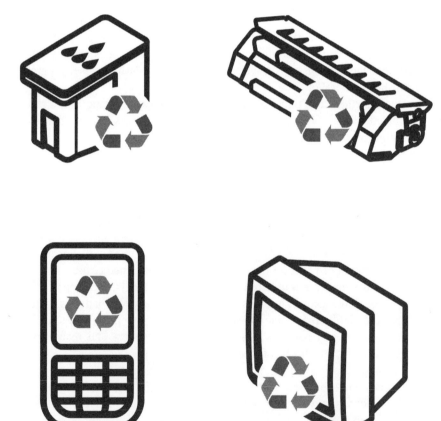

Figure 16.2.   Discarded Cartridges and Discs. By Jeff Dondero.

ALTERNATIVES TO E-WASTE LANDFILL

Donating electronics for reuse is always better than recycling. Organizations like Cell Phones for Soldiers and Verizon's HopeLine program, the Salvation Army and Goodwill, e-waste recyclers in your area, and local charities will make good use of them. Gazelle, a marketplace that can also provide some cash incentives, will help you organize a gadget drive if you're trying to raise money for your school or nonprofit organization. Enquire at local or inner-city schools about taking office equipment or post a notice on a community bulletin board offering your computer for free.

Check with your local government, schools, and universities for additional responsible recycling options. Many recyclers will pick up at your home or office. Share the Technology Computer Recycling Project provides a searchable national computer donation database to connect computer donors with charities. In the United States, the National Cristina Foundation provides computer technology to people with disabilities, students at risk, and economically disadvantaged persons.

Several retail stores, computer manufacturers, and local governments have developed recycling programs. Coupon points or discounts are available from large retailers like Staples and Office Depot toward future purchases from computers to ink cartridges.

Don't be one of millions of Americans who collectively throw away a whopping total of 1.6 billion ballpoint pens each year, according to the EPA. When the time comes to order supplies, consider refillable pens.

The following Web sites have more information on this subject: Electronic Waste Recycling Act (2003) at http://www.calrecycle.ca.gov/Electronics/ and Department Toxic Substances Control at http://www.dtsc.ca.gov/HazardousWaste/EWaste/index.cfm.

The EPA estimates that each day, Americans generate more than 3,000 tons of paper towel and tissue waste. Bring a stack of hand towels to work and hang them up in the bathrooms. Yes, you will have a little more laundry to do at the end of the week, but the trees will thank you.

17

Corporate Accountability and Responsibility

Maybe those advertisements claiming that corporations really care about people have a glimmer of truth. It's been estimated that more than 80 percent of US companies give high priority to corporate responsibility. The trend and belief in business principles today seems to surmise that companies as well as individuals have a compact with society, which states that they have a duty to act in the best interests of their communities and environments. It's called corporate social responsibly (CSR) and is aligned with sustainable responsibility in business (SRB), along with other ethical, related principles and practices.

The term "corporate social responsibility" became popular in the 1960s. It has been used loosely by some as a way to express their legal and moral consciences and by others for "greenwashing" in order to cover up or create disinformation and phony public relations to present a phony environmentally responsible front.

More accurately put, corporate social responsibility reflects a corporation's strategies, proposals, practices, and assesses and takes responsibility for the company's effects on environmental and social well-being that go beyond meeting legal regulations. It is not solely focused on maximizing profits, and calls for participation in initiatives and activities that benefit society as well as business.

Rest in peace, "old school" Milton Friedman who wrote, "There is one and only one social responsibility of business—to use its resources and engage in activities designed to increase its profits so long as it stays within the rules of the game, which is to say, engages in open and free competition without deception or fraud and to make as much money for the stockholders as pos-

Figure 17.1. Buried in Paper. By Jeff Dondero.

sible. Only people should have social responsibilities. Businesses, by their very nature, cannot."

The CSR family is joined by an inseparable link to SRB and they share and understand how acting ethically and responsibly in the marketplace can influence financial stability and the future of business, and how it is essential for entrepreneurs and executives alike.

SRB is a generally thought of as a for-profit venture that is on socially significant steroids and seeks to use its influence to create a more sustainable world—while maximizing profits. It is aimed at improving various areas ranging from environmental impact, respect, and the frugal and wise use of resources (natural and otherwise), to the social features of its local community, customers, and staff.

Both CSR and SRB require collaboration between the business sector and the local community. It's important for businesses to be aware of local customs, culture, and the details of a community's environment, its employees, as well as employee satisfaction and motivation in order to make positive contributions, minimize harmful effects, and to be a force for good in society. Both CSR and SRB aim to create a specific market space that is achieved through forming long-term partnerships and alliances with the local community and collaborating with groups such as nongovernmental and nonprofit organizations.

Corporate social responsibility should also create increasing consumer awareness about local and global issues. That includes corporate citizenship, which is the extent to which businesses are socially responsible for meeting legal, ethical, and economic responsibilities placed on them by society and their shareholders. It also calls for performance that maintains a balance between the economy and the ecosystems.

Paul Klein, founder and president of Impakt, an organization dedicated to corporate social responsibility, wrote, "The new imperative for business leaders will be to embrace the idea that the viability of their businesses depends on solving the world's most pressing societal issues." According to the International Organization for Standardization (ISO) this relationship to the society and environment in which businesses operate is "a critical factor in their ability to continue to operate effectively. It is also increasingly being used as a measure of their overall performance."

Klein suggests that one should, "Pick a big issue, declare a clear goal, and mobilize your resources. Corporations become successful because they identify problems, allocate resources to uncover and deliver solutions, and are accountable for what happens." Tyson Foods, one of the world's largest producers of meat and poultry, has a social goal of ending hunger. Through its KNOW Hunger program, Tyson donates to food banks and increases awareness of hunger issues on a large scale. At the end of 2010, Tyson had donated 78 million pounds of protein—enough to serve one meal to every American citizen.

Klein says, "Make more effort to engage the millennials in your workforce. Employees of this generation want to be rewarded in ways that go beyond monetary compensation and expect their employers to support their interest in social change. These employees also want to apply what they know to issues that they're passionate about. This means reducing or eliminating activities that have employees doing menial tasks (especially interns) that don't contribute to measurable social outcomes."

Companies demonstrate this in many ways: through waste and pollution reduction, by contributing to environmental, educational, and social programs, and by encouraging employees to volunteer for significant and philanthropic community programs. Without being cynical, large companies also understand that reporting on their sustainability goals is as much for the corporate good as for ensuring a positive image in the community and limiting their own liability should regulations put a price on carbon output or other environmental impacts.

CSR can be a key driver of innovation within an organization for everything from supply chain practices to office space operations, to how employees view their everyday work and company, to outside causes. CSR can be an incubator and innovator of ideas, encourage volunteerism, and create more internal and external value for the company.

Some of the world's largest and most profitable corporations have integrated CSR measures to promote good corporate citizenship into their mission statements and business models. A good example is the Internet gargantuan Google that has made aggressive moves to being a green giant in its efforts to use resources efficiently and support renewable power. Simply by paying attention to recycling and turning off lights, it has lowered costs and power requirements for their data centers by an average of 50 percent.

Since 1946, the retail giant Target has been committing more and more effort and assets toward local and environmental support for the communities in which it has stores and its contributions have amounted to millions each month. In the area of education alone, Target has donated close to a billion dollars since 2010.

This can mean reducing the company's carbon footprint by employing green energy, creating flexible policies for employees, installing energy-efficient lights, allowing telecommuting, encouraging the use of public transit, recycling, and supporting volunteerism and community projects.

A firm that doesn't have thousands of employees can't afford to give millions to good causes. But little deeds can pay off big time. By supporting any good cause, employees will be proud of the company and will doubly encourage clients to do business with "their" company. One payoff for creating a beneficial social culture gives employees a feeling of working for the "good guy" and contributing to something bigger than themselves and their company. People want to do things that matter; employees who participate in "giving back" are not only creating fulfillment for themselves, but for the place they call home. Encouraging this kind of involvement is an important way for companies to show job candidates, and employees, that the company's goals extend further than their product and bottom line. Candidates and workers want to know they work for a company with values that are aligned with their own, their communities, and the Earth.

One of the nation's most prized possessions is our natural capital—the stock of physical assets that include geology, soil, air, and water—and makes life (and business) possible. Corporate Ecosystem Valuation (CEV) is a concept that encourages companies to place a negative value on processes that contribute to ecosystem degradation and a positive value on the processes and decisions that lead to benefits for the environment.

The source of materials and how renewable they are is critical, as is the services they consume or support (such as fresh water or forest stewardship) and encouraging the recycling and reuse of the materials in production, the lowering of shipping costs in terms of fuel and green house gas emissions, and the use of green materials and methods in construction. The best production process is a close-looped system where the scraps from one process feed the manufacture of another. This lack of waste also contributes to net-zero carbon gain.

Companies from mom-and-pop shops to multinationals are making serious efforts to blend expense and purpose and integrate more sustainable approaches to people, the planet, and profits.

"Cause Promotion" or "Cause-Related Marketing" leverages corporate funds, in-kind contributions, or other resources to increase awareness and concern about a social issue to support fund-raising, participation, or volunteer recruitment for a cause. Corporate social marketing uses business resources to develop and/or implement a behavioral change campaign intended to improve public health, safety, the environment, or community well-being. Direct corporate philanthropy involves a company making a *direct contribution* to a charity or cause, most often in the form of cash grants, donations, and/or in-kind services.

Employee engagement activities support and encourage employees to connect with nonprofit organizations and causes. These efforts may include employees volunteering their expertise, talents, ideas, time, and even physical labor, as in Habitat for Humanity, a nonprofit, ecumenical Christian organization that builds simple, affordable housing for people in need.

Corporate support may become involved providing paid time off from work, matching services to help employees find opportunities of interest or recognition for service, and organizing teams to support specific causes the corporation has targeted. Done right, these efforts fully integrate into existing corporate social initiatives and connect the employee activities to business goals.

Another aspect of new thinking is called impact investing, a term first coined in 2007. The basic goal of impact investing is to help reduce the negative effects of business activity on the social environment, and it can be considered an extension of the CSR philosophy. Impact investing is a subset of socially responsible investing (SRI). While the definition of socially responsible investing embraces the philosophy of "do no harm," impact investing actively seeks to make a positive influence by investing in nonprofits that benefit the community, for example, as in clean technology enterprises.

A hidden asset to the bottom line is something called the "social sentiment factor." It's a measurement based on social media data that helps businesses understand how they are performing in the eyes of their consumers, what they're doing right, and how they might improve their efforts. Social sentiment indicators help companies identify trends that they can use to target new customers, develop successful marketing campaigns, create profitable products and services, and protect and improve their brand identity and image.

Social sentiment indicators are based on information users post publicly to social media like Facebook, Twitter, Yelp, blog posts, discussion groups, forums, crowd-sourced review sites about local businesses, and nonprofit organizations focused on advancing marketplace trust. If the social sentiment indicator shows a negative change in reputation, the company can address the problem through social media. By identifying those who are dissatisfied with the brand, the company can reach out, try to remedy the problem, and hopefully improve its social sentiment indicator before it grows worse. They can also make contacts with highly influential individuals with an eye to boosting their brand image in social media. Companies can also use this information to reduce the burden on their customer service e-mail and call centers by addressing questions and problems en masse via social media.

However, beware of wolves wearing sheep's corporate clothing, and who or what is "behind the curtain." Businesses can also "shill" to create a false front, threaten lawsuits to those that complain on social media, or create "Likes" with little real history or experience in their markets, and even use software to artificially increase their popularity.

Other critics claim that better governmental regulation and enforcement, rather than voluntary measures, are better alternatives to contemporary socially involvement policies. Many critics argue that effective corporate social responsibility and sustainably responsible business covenants must be voluntary as well as mandatory and that programs of social responsibility must be regulated by the government and not interfere with a corporation's rights. Others suggest that CSR and SRB programs are undertaken by companies to distract the public from questions posed by their unethical operations, unscrupulous goals, and greed, or industries that make products unhealthy to consumers and/or the environment. Furthermore, detractors questioned the "lofty" and sometimes "unrealistic expectations" in CSR and SRB goals, stating that they are merely facades to anticipate and forestall the role of governmental watchdogs over laissez-faire capitalism.

How does a person or company review and evaluate where they rank on the social responsible thermometer? Much like a resource or energy audit, an audit for corporate responsibility should be examined.

A social audit is a formal review of a company's endeavors to be socially responsible. It looks at factors such as a company's record of charitable giving, volunteer activity, energy use, transparency, environmental work, worker pay and benefits, evaluation of what kind of social and environmental impact a company is having with its products, its services, and the locations where it operates. Social audit information is optional—companies can choose whether to perform them and either release the results publicly or only use them internally. However, releasing a favorable social audit can help a company create, improve, and maintain a positive public relations image—another impact on its bottom line.

WAYS TO ENGAGE SOCIAL RESPONSIBILITY

Small businesses may think that social responsibility does not apply to them because it's the bailiwick of the "big boys." But consider the fact that large businesses only employ about 38 percent of the private sector workforce while small businesses employ 53 percent of the workforce. In fact, over 99 percent of companies are small businesses and more than 95 percent of these businesses have fewer than ten employees, according to the *Huffington Post*. So it is easy to see how small businesses have equal or more potential impact for responsibility and philanthropy in their communities.

But most companies don't have a designated corporate responsibility officer (CRO) who can handle all the aspects of what can be a lot of work and responsibility. Here are some small steps that can lead to bigger strides.

GET TO THE GRASS ROOTS

Small businesses may not have lots of employees, but they do have close ties with their employees who are familiar with their local community and can raise volunteers for their causes.

For small businesses, big donations of cash and endowments are out of the question. But they can set up an application process for small local nonprofits that helps charities in their fund-raising process. Think of inventive ways of collaborating with other small businesses, either by in-kind donations, through their Chamber of Commerce, local associations, networking groups, or just in their local community organizations. Helping each other to develop a healthy local economy, sustainable business practices, and awareness about their own natural surroundings is the kind of involvement that can help an entire community.

START YOUR OWN CAUSE MARKETING CAMPAIGN

Cause marketing for small businesses can be as simple as encouraging employees and their families to donate time or to match small cash donations, co-coordinating a community effort, posting a flier in your window, collecting spare change in store containers, giving discounts to customers who support a charity, or mentioning your favorite charity on your Web site.

Increasingly customers are expecting businesses large and small to be more philanthropic and socially responsible, so think about the fact that CSR is here to stay, and every company should consider a cohesive, integrated social strategy.

18

Building, Resource, and CSR Audit

Examining and Evaluating Output: Waste Not, Want Less

An audit can be as large or as small as a company needs or wishes it to be, revealing as much or as little about the different activities, materials, or resources it wants to measure or control. So pick the potpourri of areas or activities you wish to audit and account for and either hire a professional or DIY. Good luck.

ENERGY AUDIT DATA COLLECTION FORM

Name of Organization
Name of Contact
Position
Contact E-mail
Phone Number
Building Name
Address
Building Square Footage
Age of Building (years)
Date of Last Major Renovation
Type of Building. Major use/s and area (provide square feet):
Office
Warehouse
Retail
Other (specify)
Number of Permanent Occupants
Number of Women/Number of Men *(optional)*

Purpose of Building
Number of Floors
Daily Operational Hours (e.g., Monday–Friday 8–6; Saturday 10–4)
Days of Use per Week
Name of Utility Company
Total Number of Occupants

PLEASE CHECK ALL THAT APPLY:

This building is leased. If so:

When is the lease up for renewal (date/year)?
How long does the lease contract last (years)?

This building is owned.
The organization receives monthly bills based on accurate meter readings.
Meters are read regularly by on-site staff.
Bills are compared to monthly meter readings on a regular basis.
A Building Automation System or Energy Management Control System is in
place and is used to track utility data regularly.
The building is submetered.
The building has automated 15-minute interval or SMART meters.

HVAC ENERGY AUDIT DATA COLLECTION FORM

What type of HVAC system does the building have (e.g., constant volume,
multi-zone, VAV, etc.)?
What fuel type does this system use?
How is the HVAC system controlled (e.g., manually, DDC system, etc.)?
What are the operational set points?
What type of chilled water system does the building have, where relevant
(e.g., rotary screw chillers with cooling towers, etc.)?
How old is the chilled water system?
What is the capacity of the system?
What are the operational set points?
Do any of these systems have weather optimization sensors? If so, which
systems and what brand of sensors?
Who is responsible for managing and trouble-shooting the control system?

Are there any recurring or major occupant complaints about being too hot, too cold, etc.?

What energy efficiency efforts have been completed, started, or planned?

Are any capital improvement projects planned? If so, what are they and how will they affect the energy use of the building?

LIGHTING

Floor Name or Number
Location Description (near window, internal office, hallway, etc.)
Lamp Type
Ballast
Type
Wattage
Total Number of Lamps
Number of hours lights are left on each day
Total kWh per Day
How are lights controlled?

PLEASE SELECT WHAT IS CURRENTLY INSTALLED AT THE BUILDING:

Ground source heat pumps
Solar hot water
Commingled recycling
Lighting controls
Solar PV panels (electric)
Composting
Insulation
Wind turbines
Anaerobic digestion
Under-floor heating
Micro-hydro
Sustainable procurement
On-demand hot water heater
Geothermal
Energy-efficient windows
Weather-optimized heating sensor
Gray-water systems

Green/living roof
Low-flush toilets
Efficient HVAC systems
Rainwater harvesting
Waterless urinals
External shading
Porous pavement
Low-flow faucets
Segregated recycling
Energy-efficient lighting

OPERATIONS AND MANAGEMENT

Does the organization have an environmental policy?
Does the organization have an environmental manager?
Does the organization have an energy policy?
Does the organization review these policies on an annual basis and establish reduction targets?
Do organizational stakeholders or shareholders value environmental and social responsibility?

ANNUAL UTILITY CONSUMPTION

Building Name
Month
Electricity Usage (Btu)
Electricity (kWh)
Electricity Cost ($)
Electricity Rate ($/kWh)
Natural Gas (MMBtu)
Natural Gas Cost ($)
Natural Gas Rate ($/MMBtu)
Water (Gallons)
Water Cost ($)
Annual Totals
Total Water
Natural Gas Usage (Btu)
Total Btu
Energy Use Intensity (Btu/ft2)

BUILDING ENVELOPE

Condition (excellent, good, poor) of interior and exterior structure of walls, windows, doors, floors, roofs, garages, paved walkways and driveways, stairs, elevators, etc.

Suggestions for energy saving for any of the above.

SAVING ENERGY IN COMMERCIAL BUILDINGS

Table 18.1. Conversion Chart

1 kilowatt hour of electricity	3,412.14 Btu
1 cubic foot of natural gas	1,008 to 1,034 Btu
1 therm of natural gas	100,000 Btu
1 gallon of crude oil	138,095 Btu
1 barrel of crude oil	5,800,000 Btu
1 gallon of residual fuel oil	149,690 Btu
1 gallon of gasoline	125,000 Btu
1 gallon of diesel	129,500 Btu
1 gallon of ethanol	84,400 Btu
1 gallon of methanol	62,800 Btu
1 gallon of kerosene or light distillate oil	135,000 Btu
1 gallon of middle distillate or diesel fuel oil	138,690 Btu
1 gallon of liquefied petroleum gas (LPG)	95,475 Btu

Source: EIA.

WATER AUDIT

Agency
Facility Name
Address
Date of Assessment
Audit Coordinator(s)
Assessment Team Member

UTILITY/CONSUMPTION DATA

Water meter/s (utility meters):
Meter # Size Area serving
Meter
Monthly/Annual water consumption.

WASHROOMS

Toilets

Tank type
Number of gallons/liters per flush

Urinals

Flush tank
Number of flush tanks
Capacity of flush tank in gallons/liters
Number of urinals per tank
Flush interval

Manual flush

Number of gallons/liters per flush

Sensor flush

Number of gallons/liters per flush
Newer urinals and toilets have the gallons/liters per flush printed on the unit
  with the name of the manufacturer. Older toilets and urinals do not. If there
  is no indication on flush volume, note the manufacturer and the age of the
  toilet or urinal. With this information the manufacturer can provide flush
  volume information. The volume of the flush can be adjusted on these
  units. The flushometer manufacturer can provide the flush volume data or
  the data may be found in the building maintenance manuals.

Basins/faucets

Number of faucets. Flow rate @ gpm sensor/meter control

Showers

Number of showers. Showerhead flow rate @ gpm

KITCHENS/CAFETERIAS/LUNCH ROOMS

Number of meals served/day
Number of kitchen sinks/ faucets
Are kitchen faucets equipped with aerators?

Do spray heads have automatic shut off?
Are walk-in refrigerators/freezers water-cooled?
Are refrigerators equipped with icemakers?
Do refrigerators provide drinking water?
Are icemakers water-cooled?
Do kitchens use: garbage disposals composting neither
Is there a dishwasher? Average number of loads per week
Are only full loads washed?
Are dishes routinely pre-rinsed prior to wash?
Is frozen food routinely thawed under running water?
Are kitchen floors hosed clean? How often?
Are hoses equipped with high-pressure, water-efficient nozzles?
Number of drinking fountains: not cooled water-cooled, air-cooled
Number of vending machines/coffee makes/water coolers/ etc. connected to the domestic water system
Number of ice machines: air cooled water cooled

MECHANICAL (HVAC) EQUIPMENT

Are cooling towers in use at your facility? Number:
(Check settings for level of total dissolved solids (TDS) at blow-down and frequency.)
Is makeup water metered?
Does your municipality provide for sewer charge rebates for cooling towers?
Are you receiving a rebate?

HEATING EQUIPMENT

Are boilers in use at your facility? Number
(For steam boilers check settings for level of total dissolved solids (TDS) at blow-down and frequency.)
Is makeup water metered?

ANCILLARY EQUIPMENT

Are water-cooled air compressors in use?
Are water-cooled pumps in use?
Is making up to heating and cooling loops metered, monitored, and logged?

List any other machines that use domestic water:
Are water softeners in use? Number
Is softener regeneration automated?
If automatic regeneration, is it initiated by time meter sensor
Is resin cleaner used?

CLEANING/JANITORIAL

Are janitorial staff aware of office water conservation efforts?

PARKING AREAS

Are hoses used for cleaning parking areas?
Are hoses equipped with fine-spray/high-pressure/water-efficient nozzles?
Are dry-clean (rather than wet-clean) practices and procedures in place (i.e.,
    sweep instead of hosing, scrape before spraying, etc.)?

WALKWAYS

Are office sidewalks and outside walls pressure-washed on a regular basis?
When are fountains running?
Do fountains use recycled water?
Are they part of a closed-loop system?

PLANNED ACTION

In this section, planned actions to reduce water consumption should be listed.
For example, if your building is using about 3 gallons/20 liters per flush toilet, the action may be to change those for low-flush toilets (6 liters per flush).

BACKGROUND

Baseline water usage
Local water provider
Where does your water come from?
Number of buildings at facility. Size of buildings (area)
Area of grounds. Size of motor pool (# of vehicles)

Number of employees per shift. Number of shifts per day
Average number of visitors/occupants per day (if applicable)
Water pressure at your facility (psi)
(Often reducing water pressure by merely 10 or 15 percent can reduce water consumption significantly without interfering with daily consumption activities. Water pressure that is too high can result in leaks.)
Does staff have good general water conservation awareness habits?
Have you ever had a water-balance or leak-check?
Is there an on-site wastewater treatment facility?
If so, give a brief description of the facility, flow rates, chemical additions, and average cost (per unit volume).
Size, type, and location of water meter(s)
Describe the amount, location, and use of untreated groundwater if you use it at your facility.
What, if any, water efficiency measures are already in place?

PERSONAL WATER CONSUMPTION

Number of restrooms
Number of toilets (total) Note: Many fixtures have the average flow rate printed on the fixture itself, along with the make and model. If you cannot find this printed information, consult your maintenance staff or facility manager.
Type of toilets and average water consumption in gallons per flush (gpf)
Note: Most toilets are either gravity flush, flush valve/flushometer/tankless, or pressurized tank types.
Toilet type. Number. Average gpf
Are toilets equipped with toilet dams or low-flow flapper valves?
Do flush valve (tankless) toilets have water-saving diaphragms?
Are toilets equipped with automatic water-flushing systems?
If so, what is the timing cycle?
Are the sensors/timers coordinated with regular work hours?
Total water consumption per workday from toilet flushes, assuming each employee/occupant uses the bathroom 4 times per workday
Number of restroom faucets (total). Faucet flow rate @ gpm
Are faucets equipped with aerators?
Are faucets equipped with automatic or metered shutoff mechanisms?
Number of showers (total). Showerhead flow rate gpm
Number of drinking fountains. Fountain flow gpm
Are fountains air-cooled or water-cooled?

KITCHENS

Number of staff kitchen areas
Number of kitchen sinks/faucets. Faucet flow rate gpm
Are kitchen faucets equipped with aerators?
Do refrigerators use water coolant systems?
Are refrigerators equipped with icemakers?
Are icemakers water-cooled or air-cooled
Do refrigerators provide drinking water?
Do kitchens use garbage disposals, composting, neither
Is there a dishwasher?
Average number of loads per week

CAFETERIAS

Number of kitchen areas
Number of faucets (total). Faucet flow rate @ gpm
Are faucets equipped with aerators?
Are faucets equipped with flow restrictors?
Number of dishwashers. Make & Model
Average number of loads per day. Water consumption per load:
Are dishes pre-washed?
Is potable water used for pre-washing dishes?
Is dishwasher wastewater reused?
Does the flow of water to the dishwasher stop when the flow of items being washed stops?
Does the flow of water to the garbage disposal stop when the disposal motor stops?
Number of ice machines. Are ice machines air-cooled or water-cooled?
Number of refrigerators
Do refrigerators use water coolant systems?
Is frozen food routinely thawed under running water?
Do steam tables on buffet lines use potable water?
What is done with water from steam tables at closing time?
Are kitchen floors hosed clean?
Are hoses equipped with high-pressure, water efficient nozzles?
Are linens washed on-site?

LABORATORY CONSUMPTION

Number of labs (total in facility)
Number of sinks/faucets and flow rate @gpm
Are faucets equipped with aerators?
List lab equipment that uses water in any way
Equipment amount used closed-loop? Potable? Reused?
Describe lab procedural/clean-up practices that consume water. Are procedures and clean-up practices posted in the lab?

MECHANICAL CONSUMPTION

Number of water heater(s). Size
Are water softeners in use?
Backwash? Frequency & Duration
Are drink machines in vending areas air-cooled water-cooled?
Are cooling towers in use at your facility?
For each cooling tower, approximate how much makeup water is needed or used to replace water lost to blow-down, evaporation, and other process inefficiencies.
Check settings for level of total dissolved solids (TDS) at blow-down and frequency.
Are boilers in use at your facility?
For each boiler, approximate how much makeup water is needed or used to replace water lost to blow-down, evaporation, and other process inefficiencies.
Check settings for level of total dissolved solids (TDS) at blow-down and frequency.
Are water-cooled air compressors in use?
Are water-cooled pumps in use?
List any other machines that use non-contact cooling water:

HEATING, VENTILATING, AND
AIR CONDITIONING (HVAC) CONSUMPTION

What type of HVAC system do you have?
Does your HVAC system have condensate collection and/or re-use?
Is your HVAC system always on?
Is your HVAC system air-cooled or water-cooled?
If water-cooled, is your system open-loop or closed-loop?

CLEANING USE

Motor Pool:
Number of automobiles. Where are they washed? How frequently?
Number of watercraft? Where are they washed? How frequently?
Are hoses used?
Are hoses equipped with fine-spray/high-pressure/water-efficient nozzles?
Are dry-clean (rather than wet-clean) practices and procedures in place? (i.e.,
sweep instead of hosing, scrape before spraying, etc.)
Are office windows washed on a regular basis?
How often?
Are office sidewalks and outside walls pressure-washed on a regular basis?
How often?

JANITORIAL USE

Are janitorial staff aware of office water conservation efforts?
Are there areas that janitors mop?
Area mopped (square feet)
How often?
Are hoses used?
Are hoses equipped with fine-spray/high-pressure/water-efficient nozzles?
Are dry-clean (rather than wet-clean) practices and procedures in place? (i.e.,
sweep instead of hosing, scrape before spraying, etc.)
List other janitorial practices that consume water.

LANDSCAPING CONSUMPTION

What types of vegetation are planted in the landscaping surrounding your
facility?
Plant Name Native? Average water consumption
Soil type (in your region). Does your landscape use mulch?
Does your facility have an irrigation system? Types
Where does the system irrigate?
How often? Summer/Winter
Time of day? Summer/Winter
Is there a rain gauge incorporated in your system?
Are there manual override controls for your system?
Are hoses used for irrigation?
Are hoses equipped with fine-spray/high-pressure/water-efficient nozzles?

Does your facility have any pools or fountains?
When are fountains running?
Typical water consumption @ gpm
Do fountains use recycled water?
Are they part of a closed-loop system?
Are paved areas swept clean, blown clean, or hosed?

MAINTENANCE

Are faucets, pipes, and plumbing checked regularly for leaks?
How often?
Is there regularly scheduled preventive maintenance in your facility?
Is maintenance documented with standard records or inspection logs?
If you contract with a maintenance company, how quickly does maintenance staff respond and repair leaks?
If you control your own maintenance program, how do you handle reporting and repair of leaks?
How quickly are leaks usually repaired?

CONTRACTOR WATER USE

Are contractors aware of agency water conservation efforts?
Do any contractors have access to or use of facility water?
Do contractors know where the water line is located?
Notes/ Other Water Consumption

A CSR AUDIT

A social audit is a formal review of a company's endeavors in social responsibility. A social audit looks at factors such as a company's track record of charitable giving, volunteer activity, energy use, transparency, worker pay, scheduling, benefits, and environment to examine and evaluate the kind of social and environmental impact a company is having in the locations where it operates.

The auditing process may be conducted internally by your company. However, you can choose to have one conducted by an outside consultant who will impose minimal biases, which may prove to be more beneficial to your company. Consider the fact that, as with a financial audit, an outside auditor brings credibility to the evaluation. This credibility is essential if management is to take the results seriously and if the general public is to believe in your company's public relations, social cause activities, and social cause marketing.

Once the social responsibility audit is complete, it may be distributed internally, or both internally and externally, depending on your company's goals and findings. Some companies publish a separate periodic report on their social initiatives and later have it available on their Web site. And nearly all publicly traded companies now include a section in their annual report devoted to social responsibility activities.

A CSR AUDIT SHOULD COVER THESE QUESTIONS AND TOPICS

People

Fundamental human rights, freedom of association, freedom of expression.

Business Behavior

Relations with clients, suppliers and subcontractors, prevention of corruption and anti-competitive practices.

Human Resources

Labor relations, working conditions, equitable salaries, health and safety, career development and training, collective bargaining, nondiscrimination in hiring and community affairs.

Environment

Incorporation of environmental considerations into the manufacturing and distribution of products, and the use, recycling, and disposal of waste.

Community Involvement

Impacts on local communities, contribution to social and economic development, general community interests.

THE CORPORATE SOCIAL RESPONSIBILITY (CSR) PLAN

Who is going to be the person in charge of creating and launching the CSR program(s)?

What is your mission statement and how does it align with your responsibilities to your community and stakeholders?

Does your business plan reflect what the company sees as its corporate social responsibilities?

What were specific wins and losses—where, what, and why?

Have board and management agreed on and communicated an explicit commitment to CSR?

Do board and management share a common definition of CSR as it relates to the company, its sector, and broader societal trends?

Has the board developed a common understanding of the company's business case for CSR?

Have board and management developed a CSR vision for the company?

Is CSR incorporated into the company's mission, vision, and values?

Is the company Code of Conduct/Ethics incorporated into CSR?

Is the board aware of CSR issues specific to the company and community?

Has the company developed a formal strategic process that will help measure the company's actual social performance against the social objectives it has set for itself?

How have decision making, mission statement, guiding principles, and business conduct aligned with social responsibilities?

Has the audit examined and evaluated the interests and objectives of the employees and stakeholders?

Note: A CSR audit can be used to examine the company's vulnerabilities, then to decide how to launch new social initiatives inside or outside the company. These initiatives can aid in capturing market share from direct competitors, and can even help introduce new products.

Code and Conduct

Do you have a corporate code of conduct?

How do you share and communicate your corporate code of conduct with your employees and community?

How are companies in other industries complying with their social responsibilities that are similar to yours?

Ideas and Breakthroughs

Based on the findings and discovery above, what can you do differently to strategically set your company apart?

What do you want to do better? What things need improvement at your company?

What do you want to continue doing? What area of your community, or who in the community, is most affected by your business?
What community issues are likely to affect your business and/or employees?
What role does your company want to play in the community?
What have you done in the past to help this relationship?
What are your employees' social interests and to what are they committed?
Does your business plan and business strategy fit well with your CSR?

Assessments

Consider each one of these alternatives: Sponsorship, co-branding, licensing, new product promotion, philanthropic investments, donations of products and services, and employee involvement.
Which one(s) seem best for your situation analysis? Why?
What does your corporate leadership team think? What do your employees think? How are you going to evaluate and measure the outcomes for your social cause initiatives?
What kind of resources are you willing to commit to CSR and for how long?
What criteria will be used for judging the success of the CSR plan?
How do you motivate employees in regard to CSR?
How will the method of selecting CSR plans be employed?
Which social cause initiative looks like the best for your situation? Why?
Define and state your deadlines to get started with a CSR plan.
How are the employees going to participate in the CSR program(s)?
Are you going to share your audit internally and/or externally?
Will employees and shareholders examine and express their opinions of the CSR plan?
Who will prepare a summary of your corporate social responsibility audit?

Action

How will you plan who will be in charge of what aspects of the CSR plan? Cut the plan into sections.
How will you form committees for various aspects of the plan?
How will you motivate employees?
How will you get the community involved?
How will you initiate the plan?
How will you follow through with the plan?
How will you finish the plan?
How will you evaluate the execution of the plan?
Information provided by http://business.edf.org/files/2014/03/Sample-Water-Audit-Forms.pdf and http://www.nrel.gov/docs/fy11osti/50121.pdf

19

Need Some Help?
Agencies and Organizations to Aid in Conservation

The agencies and organizations listed below are a partial directory that can be of aid in answering questions, providing resources, and memberships to enhance the quest for all manner of resource information and expertise. We suggest that you look in various places from community libraries, chambers of commerce, regional clubs and associations, police and fire departments, community health agencies, schools, city halls and county seats, and, of course the Internet for resources, especially local ones from nonprofits to parks and recreation departments.

CHECK THE INTERNET

Sustainability organizations seek to implement strategies, which provide them with economic and cultural benefits attained through corporate responsibility including: ecological motives, economic motives, and legal and regulatory pressures. Look for a list of sustainability organizations by topic (business, law and policy, news, nonprofit, etc.) as well as by region (international, national, state, local).

ENERGY-EFFICIENCY ORGANIZATIONS AND ASSOCIATIONS

Efficiency$mart: Nonprofit dedicated to advancing energy efficiencies for economic prosperity and environmental protection. www.efficiencysmart .org/news-events/news/120

Consortium for Energy Efficiency (CEE): US and Canadian consortium of gas and electric efficiency program administrators working together to accelerate the development and availability of products and services, using the power of the mass markets to advance energy-efficient technologies that benefit customers and the environment: https://www.cee1.org/

Database of State Incentives for Renewables and Efficiency (DSIRE): Comprehensive resource of incentive programs and tax credits, and exemptions for energy efficiency and renewables. www.dsireusa.org/

Illuminating Engineering Society of North America: Recognized technical authority on illumination, providing information on all aspects of good lighting practices, programs, publications, and services. www.ies.org/

Lighting Research Center: Part of the School of Architecture, Rensselaer Polytechnic Institute, and the leading university-based lighting research center with an international reputation as an objective source. Provides information about lighting technologies, applications, and products. http://www.lrc.rpi.edu/

PennFuture Energy Center: Provides information and resources about energy efficiency and ways to lower electric costs. Until recently, much of the technology and information about how to waste less money and use energy more wisely was beyond the reach of most families and small businesses, but thanks to the Energy Savings Law (Act 129) passed in late 2008, Pennsylvania's electricity companies are now required to offer programs free of charge to customers of all types—residential, commercial, nonprofit/government, schools, and industrial—that will help them reduce their electricity use through energy efficiency. http://sustainableferc.org/partners/penn-future-energy-center/

Small Business Administration: Offers small business owners advice, tips, and tools for becoming more energy-efficient. https://www.sba.gov/category/navigation-structure/energy-efficiency

US Department of Energy, Office of Energy Efficiency and Renewable Energy: Provides educational resources, analysis tools, information about their latest projects, and links to industry information and job opportunities. http://energy.gov/eere/office-energy-efficiency-renewable-energy

EnergyStar for Business: The buildings where we work, shop, play, and learn account for nearly half the nation's energy use. Explore the EnergyStar site and find tools and resources to help businesses and organizations save energy and reduce their carbon footprint. https://www.energystar.gov/buildings/facility-owners-and-managers/small-biz

EPA Center for Corporate Climate Leadership: Serves as a virtual resource center to encourage companies to identify and achieve cost-effective greenhouse gas emission reductions and to build their climate leadership activities into their supply chains and beyond. https://www.epa.gov/climateleadership

EPA WaterSense: Helps assess water-efficiency. Incorporating water-efficiency programs is an effective way for businesses to reduce operating costs by saving on electric power, gas, chemical, and wastewater disposal expenses. By employing water-efficient practices, you can convey an image of stewardship to employees, customers, and the general public because you are helping to conserve water resources for future generations. https://www3 .epa.gov/watersense/test_your_watersense.html

EPA Toxics Release Inventory (TRI) Program: Publicly available EPA database that contains information on toxic chemical releases and waste management activities reported annually by certain industries as well as federal facilities. https://www.epa.gov/toxics-release-inventory-tri-program

EPA Safer Choice Program (formerly Design for the Environment): Uses chemical assessment tools and expertise to create safer and more efficient chemical materials, processes, and technologies. Safer Choice focuses on industries that combine the potential for chemical risk reduction and improvements in energy efficiency with a strong motivation to make lasting, positive changes. https://www.epa.gov/saferchoice

EPA Risk Management Plan (RMP) Rule: The Clean Air Act requires facilities that produce, handle, process, distribute, or store certain chemicals to develop a risk management program, prepare a Risk Management Plan (RMP), and submit the RMP to EPA. Learn how to prepare and submit an RMP, and how to access and review RMP information. https://www.epa.gov/rmp

EPA Superfund: Provides information for businesses: legislation and regulations, compliance and enforcement, databases and software, contracts and more. https://www.epa.gov/superfund

EPA Phaseout of Ozone-Depleting Substances: Provides information for companies that use or manufacture ozone-depleting substances and their substitutes, including information about the chlorofluorocarbon phase out (a class of compounds of carbon, hydrogen, chlorine, and fluorine, typically gases used in refrigerants and aerosol propellants that harmful to the ozone layer in the Earth's atmosphere). https://www.epa.gov/ods-phaseout

American Council for an Energy-Efficient Economy (ACEEE): Sponsors an international conference on energy efficiency in buildings. Publishes reports on technical and policy issues relating to energy efficiency. http://aceee.org/

Association of Energy Conservation Professionals (AECP): Nonprofit energy education and advocacy organization aiming to provide, promote, and advocate for energy conservation. http://www.aecpes.org/

Best Practices Energy Smart Technologies for Today: An initiative of DOE's Office of Industrial Technologies (OIT) to help manufacturing plants improve energy-efficiency and reduce costs in motor, steam, compressed air, and process heat systems. Works with its allied partners to assist thousands of industrial end-users nationwide achieve increased energy efficiency and productivity improvements. www.nrel.gov/docs/fy99osti/25972.pdf

Center for Energy Efficiency and Renewable Energy (CEERE): Provides technological and economic solutions to environmental problems resulting from energy production, industrial, manufacturing, and commercial activities, and land use practices. https://mie.umass.edu/center-energy-efficiency -and-renewable-energy

Center for Resource Solutions: Promotes clean and efficient energy through six programs that focus on national and international renewable energy issues; encourages the transfer of sustainable technologies and the development of sustainable energy practices. https://resource-solutions.org/

Clean Energy States Alliance (CESA): Eleven states across the United States have established funds to promote renewable energy and clean energy technologies. CESA provides information and technical services on these funds to build and expand clean energy markets in the United States including publications, meetings, programs, and membership information. http://cesa.org/

Energy and Energy Conservation: Lists recent releases of books on energy topic issues, including order page and special offers. https://www.nap.edu/ topic/283/energy-and-energy-conservation

Energy Conservation, National Wildlife Federation: Concerned about global warming, the site provides advice on how to reduce energy consumption. https://www.nwf.org/How-to-Help/Live-Green/Energy-Conservation.aspx

Energy Conservation News and Resources: This is mostly a collection of links to different resources. http://www.energyconservationinfo.org/

Department of Energy, US Energy Information Administration (EIA): Provides efficiency indicators and measurements of greenhouse gas as related to energy use and energy efficiency; also provides a list of links regarding energy savings. http://www.eia.gov/

Energy Technologies Area (ETA), Lawrence Berkeley National Laboratory: Conducts research and development leading to better energy technologies and reduction of adverse energy-related environmental impacts. https:// eetd.lbl.gov/

Federal Energy Management Program (FEMP): Helps federal agencies reduce building energy costs, increase energy efficiency, use renewable energy, and conserve water. https://energy.gov/eere/femp/about-federal-energy -management-program

Landscaping for Energy Conservation: Promotes landscaping as a way of reducing the amount of energy used. http://extension.colostate.edu/topic -areas/yard-garden/landscaping-for-energy-conservation-7-225/

Midwest Renewable Energy Association: Nonprofit organization that promotes a sustainable future through renewable energy and energy efficiency. https://www.midwestrenew.org/

Northeast Energy Efficiency Partnerships (NEEP): Regional nonprofit that increases and coordinates energy-efficiency and market transformation efforts in the US Northeast. http://www.neep.org/

Northeast Sustainable Energy Association: Promotes the understanding, development, and adoption of energy conservation and nonpolluting, renewable energy technologies. Programs and activities focus on the northeastern United States (from Washington, DC to Maine). http://nesea.org/

Northwest Energy Efficiency Alliance (NEEA): Nonprofit consortium in the US Pacific Northwest works to transform markets for energy-efficient products and services. Online reports on products and technologies for industry, agriculture, homes, and other applications are available. http://neea.org/

Northwest Energy Efficiency Council, Lighting Design Lab: An energy-efficient lighting design resource for the northwest United States that offers training classes, demonstrations, consultations, and related services. www.neec.net

Oak Ridge National Laboratory (ORNL), Energy-Efficiency and Renewable Energy Program: Conducts research and development on technologies to reduce energy consumption in homes, industry, transportation, and utilities. Advanced materials, improved insulation, telecommuting, and improved heat pumps. http://web.ornl.gov/sci/eere/

What You Need to Know About Energy: Provides information on energy efficiency, cost and sources. Download a PDF of "What You Need to Know About Energy" by the National Research Council for free. Description: American society, with a standard of living. https://www.nap.edu/catalog/12204/

BUILDING

By using more efficient building methods and materials, it is estimated that we could reduce the energy, resource consumption and/or waste production by 50 to 60 percent without decreasing value, aesthetics, or function of structures, and while taking the Earth's finite resources and natural environment into consideration.

GREEN BUILDING ORGANIZATIONS

- BuildingGreen provides authoritative information on environmentally responsible building design and construction from the publishers of *Environmental Building News*.

- Oikos provides detailed information on sustainable design and construction.
- Healthy Building Network promotes healthy building materials as a means of improving public health and preserving the global environment.
- Northwest EcoBuilding Guild is an association of building professionals and homeowners interested in ecologically sustainable building.
- Operation Fresh Start is designed to empower communities as they recover from floods, hurricanes, earthquakes, and other natural disasters by providing tools and resources to rebuild homes and businesses using sustainable principles and technologies.
- US Green Building Council is the nation's foremost coalition of leaders from across the building industry working to promote buildings that are environmentally responsible, profitable, and healthy places to live and work.
- World Green Building Council is a US nonprofit corporation striving to help countries form national green building councils of their own by providing valuable information and resources to its affiliates and members.
- ThePOOSH.org is an Estonian-based NGO with an international scope that connects volunteers with sustainable build projects all around the world to exchange labor, skills, and knowledge.

SUSTAINABLE BUSINESS ORGANIZATIONS

Sustainable business organizations participate in environmentally friendly or green practices in order to make certain that all processes, products, and manufacturing activities sufficiently address current environmental concerns while still retaining a profit. There are many organizations and networks currently interacting with businesses in order to integrate sustainability into their central goals and contribute to the environmentally and socially responsible business movement.

Business and Industry Resource Venture provides free information, assistance, and referrals to help Seattle businesses improve their environmental performance.

Business for Social Responsibility is a business membership organization.

Center for Sustainable Economy is a nonpartisan research and policy organization that promotes innovative tax and other market-based approaches to achieving a sustainable economy.

Ceres is a national coalition of environmental, investor, and advocacy groups working together for sustainable prosperity. These groups form a community of forward-looking companies that have committed to continuous environmental improvement by endorsing Ceres principles, a ten-point code of environmental conduct.

Cool Companies is a guide for businesses seeking to cut energy costs and reduce pollution.

Future 500 brings together strategically selected companies, NGOs, and opinion leaders to overcome mutual distrust and advance systemic solutions.

Global Environmental Management Initiative (GEMI) helps set the standard for reporting environmental, economic, and social activity and is the global leader in developing insights, networking, and creating collaborative sustainability solutions for business.

GreenBiz is a resource center on business, the environment, and the bottom line.

Lifestyles of Health and Sustainability (LOHAS) is a market segment focused on health and fitness, the environment, personal development, sustainable living, and social justice.

Natural Capital Institute researches and initiates projects relating to the relationship between human and living systems, with particular emphasis on natural capital, green business, biomimicry, innovative design, and social justice. Particular focus on Socially Responsible Investing Project and Database, and Wind Energy Resources Project.

Net Impact is a network of emerging business leaders committed to using the power of business to create a better world. It is also the most progressive and influential network of MBAs in existence today. Originally founded as Students for Responsible Business in 1993, Net Impact serves to broaden business education, refine leadership skills, and pursue professional goals while building networks.

COMMUNITIES

Sustainable community organizations often encourage and cultivate collaborative community projects and education programs that improve connections between businesses, institutions, and the public with their communities, the natural environment, and each other. For example, the Sustainable Community Initiatives (SCI) organization:

- Acts to help the business community, local government, and the general public develop an awareness of the value of sustainable community development.

- Endeavors to combine knowledge (through public education) and action (in the form of community programs) to promote prosperous and healthy communities.
- Explores opportunities to develop educational, entrepreneurial, and environmentally sound community-based projects.
- Develops materials on the efficient use of urban environmental resources and sustainable activities for distribution to local businesses and to the general public.
- Conducts sustainability seminars and workshops for a variety of audiences, for example, financial institutions, housing development organizations, continuing education programs, and community organizations.
- Participates in efforts to promote more sustainable tax policies and land use codes.
- Integrates into all of SCI's undertakings, the fundamental elements of community sustainability, community partnership, community enterprise, community conservation. and community design.

ENVIRONMENTAL SUSTAINABILITY ORGANIZATIONS

Bioneers is working to preserve biological and cultural diversity.

Center for the Advancement of the Steady State Economy. Their mission is to advance the steady state economy, with stabilized population and consumption, as a policy goal with widespread public support. They educate about the conflict between economic growth and environmental protection, ecological and economic sustainability, and national security and international stability.

Center for a New American Dream helps Americans consume responsibly to protect the environment, enhance quality of life, and promote social justice.

Climate Solutions promotes a regional approach to global warming solutions.

Friends of the Earth (FOE) is an international environmental organization dedicated to preserving the health and diversity of the planet for future generations.

International Institute for Environment and Development (IIED) promotes sustainable patterns of world development through collaborative research, policy studies, networking, and knowledge dissemination.

International Union for the Conservation of Nature (IUCN) seeks to conserve the integrity and diversity of nature and to ensure that any use of natural resources is equitable and ecologically sustainable.

Northwest Earth Institute is a pioneer in taking Earth-centered education programs to people where they spend their time—in their neighborhoods, workplaces, homes, schools, and centers of faith.

Renewable Energy Policy Project (REPP) supports the advancement of renewable energy technology through policy research.

Rocky Mountain Institute works extensively with the private sector, as well as with civil society and government, to create abundance by design and to apply the framework of natural capitalism.

Sightline Institute is an independent, nonprofit research and communications center—a think tank—founded by Alan Durning in 1993.

Union of Concerned Scientists (UCS) combines independent scientific research and citizen action to develop innovative, practical solutions and to secure responsible changes in government policy, corporate practices, and consumer choices.

Zero Emissions Research and Initiatives (ZERI) is a global network of creative minds seeking solutions to world challenges. The common vision shared by the members of the ZERI family is to view waste as resource and seek solutions using nature's design principles as inspiration.

Zero Waste Alliance is a national leader providing assistance to industry sectors and organizations for development and implementation of standards, tools, and practices that lead to a more sustainable future through the reduction and elimination of waste and toxics.

LAW AND POLICY SUSTAINABILITY ORGANIZATIONS

Earth Policy Institute is dedicated to building an environmentally sustainable economy; raising awareness to support public response to population growth, rising CO_2 levels, loss of species, and other trends that are affecting the Earth.

Global Exchange is nonprofit organization that envisions a people-centered globalization that values the rights of workers and the health of the planet; prioritizes international collaboration as central to ensuring peace; and aims to create a local, green economy designed to embrace the diversity of our communities.

Apollo Alliance is a coalition of labor, business, environmental, and community leaders working to catalyze a clean energy revolution that will put millions of Americans to work in a new generation of high-quality, green-collar jobs. Inspired by the Apollo space program, this organization promotes investments in energy efficiency, clean power, mass transit, next-generation vehicles, and emerging technology, as well as in education and training.

Institute for Local Self-Reliance is a comprehensive resource for policy-makers, organizations, and activists looking for innovative public policies enacted around the world that can be used to make communities vibrant and strong. It proposes a "new set of rules," which supports humanly scaled politics and economics.

GOT A SCOOP? GO TO THE
NEWS SUSTAINABILITY ORGANIZATIONS

EcoSnoop is a photo-driven system with the objective of identifying waste and making people aware of ways to become greener.

Worldchanging is a nonprofit media organization headquartered in Seattle, Washington, that comprises a global network of independent journalists, designers, and thinkers. It covers the world's most innovative solutions to the planet's problems, and inspires readers around the world with stories of new tools, models, and ideas for building a bright green future.

Worldwatch Institute is dedicated to fostering the evolution of an environmentally sustainable society—one in which human needs are met in ways that do not threaten the health of the natural environment or the prospects of future generations.

TriplePundit is an online publication dedicated to promoting the concepts of sustainability to a business audience. The name is a reference to the "Triple Bottom Line" which seeks to balance economy (profit), society (people), and the environment (the planet)—a key tenet to sustainable thinking.

NONPROFIT SUSTAINABILITY ORGANIZATIONS

Corporate Watch is a nonprofit organization committed to "holding corporations accountable." It functions as a resource on corporate globalization research, news stories, and ways to take action. It also organizes campaigns against political campaign financing, green-washing, and war-profiteering.

Earth Island Institute works to find solutions to environmental problems by developing and supporting projects that promote the conservation, preservation, and restoration of the Earth.

Environmental Defense Fund is a leading national nonprofit organization representing more than 700,000 members. Guided by science, the Environmental Defense Fund evaluates environmental problems and works to create and advocate solutions that win lasting political, economic, and social support.

Environmental Working Group (EWG) is a not-for-profit environmental research organization that uses the power of information to improve public health and protect the environment by reducing pollution in air, water, and food. Based in Washington, DC, and with an office in Oakland, California,

EWG conducts groundbreaking, computer-assisted research on a variety of environmental issues.

Green Cross International is a nonprofit NGO founded by Mikhail Gorbachev with a mission to help create a sustainable future by cultivating harmonious relationships between humans and the environment.

Social Venture Network (SVN) promotes new models and leadership for socially and environmentally sustainable business in the twenty-first century. As a nonprofit network, it does this by championing member initiatives, information services, and community forums.

Fair Trade USA is a nonprofit organization, one of twenty members of Fairtrade Labeling Organizations International (FLO), and the only third-party certifier of Fair Trade products in the United States.

US FEDERAL GOVERNMENT AND RELATED AGENCIES

Energy Efficiency and Renewable Energy Network (EREN) is the US Department of Energy's comprehensive resource for energy efficiency and renewable energy information. The site provides a wealth of information for sector energy efficiency programs, such as those for the building and industrial sectors. Technologies are presented as well as featured sites and specialized resources.

ENERGY TAX INCENTIVES

If you purchase energy-efficient appliances or make energy-saving improvements to your home or business, you can save money on your utility bills. You can also save more money on these purchases, in the form of tax incentives, such as tax credits and rebates, or sales tax holidays. Use these databases to find out if you qualify for state, local, and federal incentives:

- Database of State Incentives for Renewables and Efficiency (DSIRE)
- Department of Energy (DOE): Tax Credits, Rebates, and Savings
- Offers and Rebates from EnergyStar Partners

FEDERAL TAX CREDITS FOR ENERGY EFFICIENCY

A tax credit can provide significant savings. It reduces the amount of income tax you have to pay. Unlike a deduction, which reduces the amount of income subject to tax, a tax credit directly reduces the tax itself. The IRS has provided the following guidance regarding the tax credits for constructing energy-efficient new homes available under the Energy Policy Act of 2005:

- Tax Credits for Home Builders. Eligible contractors need to fill out IRS Form 8908 to get the tax credit.
- The Tax Relief and Job Creation Act of 2010 modifies and extends the energy-efficient appliance credit for certain dishwashers, clothes washers, and refrigerators manufactured after December 31, 2010.
- Manufacturer's Energy-Efficient Appliance Credit and the associated Form 8909.
- Tax Deductions for Commercial Buildings.
- Tax Incentives for Hybrid, Electric, and Alternative Fuel Vehicles.

ENVIRONMENTAL GRANTS AND LOANS

Use the links below to help find financial resources to pay for energy-efficient upgrades to your facilities, to finance your business's innovative environmental products and technologies, and to support your environmentally friendly business.

Search grants.gov by category to see grant applications related to energy, environmental topics, natural resources, transportation, science and technology, agriculture, and disaster prevention and relief. The Partnership for Sustainable Communities grants site can also help you identify relevant grants on grants.gov.

The Environmental Protection Agency (EPA) has several grant programs, many of which are listed on this site. In addition, links to other specific programs are provided.

- Brownfields Grants for redeveloping formerly polluted real estate
- Green Building Grants
- Integrated Pest Management Grants
- National Center for Environmental Research
- National Science Foundation (NSF) Grants
- Environmental Engineering
- Environmental Sustainability
- Energy for Sustainability

SMALL BUSINESS INNOVATION GRANTS

A few federal programs provide grants to small firms engaged in scientific research and development (R&D). The federal government's SBIR (Small Business Innovation Research) and STTR (Small Business Technology Transfer) programs award a specific percentage of federal R&D funds to qualified small businesses. SBIR/STTR programs encourage small firms to undertake scientific research that helps meet federal R&D objectives, and have high potential for commercialization if successful.

US DEPARTMENT OF ENERGY TAX, CREDITS, REBATES, AND SAVINGS LIBRARY

Browse this listing of financing opportunities to find programs that your business is eligible for:

EnergyStar Building Upgrade Manual, Financing Section: Learn about the wide-range financing options available to businesses. Financing is available from small improvements to complete system upgrades. Additional guidance is also available here.

EnergyStar Rebate and Special Deals Finder: Use this online tool to find special offers and rebates on office equipment, electronics, appliances, and lighting products in your local area. https://www.energystar.gov/index.cfm?fuseaction=rebate.rebate_locator

Industrial Technologies Program (ITP): Lead government program working to increase the energy efficiency of US industry. Along with their partners, ITP helps research, develop, and deploy innovative technologies that companies can use to improve their energy productivity, reduce carbon emissions, and gain a competitive edge.

Industrial Assessment Centers (IACs): Provides eligible small- and medium-sized manufacturers with no-cost energy assessments. Sponsored by the US Department of Energy, Industrial Technologies Program.

Manufacturing Extension Partnership (MEP): Provides customized technical assistance programs to manufacturers in the areas of process improvement, supply chain management, and business operations. Programs include assistance with implementing energy-saving measures in the manufacturing process, sponsored by the National Institute of Standards and Technology.

Index

About the Author

Jeff Dondero has a diverse background and experience in writing, ranging from web content, B2B, books, hard news, to interviews to feature writing. He began his career as a stringer and freelancer for the *San Francisco Examiner*, worked as a reporter and editor for several suburban newspapers, was the entertainment editor for *The Marin Independent Journal*, was a writer and editor of various magazines, wrote for KTVU-TV in the San Francisco Bay Area, toiled in a trade magazine mill, and created a website dedicated to sustainable construction industries (http://www.greenbuildingdigest.net/). He was invited to be a writer-in-residence at an art colony in Rancho Vista, Arizona, in 2014, where he wrote a slim volume of poetry. He continues to expand his national readership with books, social media, various writers' blogs and websites, and radio and television appearances.